Effectively Using The OSCILLOSCOPE

by

Robert G. Middleton

A Revision of

101 Ways To Use Your Oscilloscope

by Robert G. Middleton

Howard W. Sams & Co., Inc.
4300 WEST 62ND ST. INDIANAPOLIS, INDIANA 46268 USA

THIRD EDITION
SECOND PRINTING—1982

International Standard Book Number: 0-672-21794-5
Library of Congress Catalog Card Number: 80-54462

Printed in the United States of America.

Contents

SECTION 3

Television Tests and Measurements

SECTION 4

Digital Logic Tests

SECTION 5

Semiconductor Tests and Measurements

SECTION 6

Miscellaneous Applications

SECTION 7

Oscilloscope Performance Checks

Introduction

In recent years electronics has undergone significant changes—some technicians would say revolutionary changes. For example, the vacuum tube has almost disappeared from the scene (with the exception of the cathode-ray tube). The bipolar transistor and field-effect transistor are now being extensively replaced by integrated circuits. Today, an integrated-circuit operational amplifier could be confused with an older type of transistor, except for its more complex basing. Silicon controlled rectifiers and related devices have made extensive inroads into the power-type bipolar transistors formerly used in television receivers. More dramatically, a digital revolution has occurred, and receiver tuning in particular is often accomplished by computer-type arrangements. Specialized television receivers provide digital-controlled readouts of the operating channel and the time, with provision for extensive preprogramming of receiver operation. Digital filtering techniques are currently used for optimum separation of chroma and Y signals.

Evolution of the oscilloscope has kept pace with the rapid advance of electronics technology. Not too long ago the service-type oscilloscope illustrated in Fig. I-1 was unheard of, although its predecessor was used in laboratories. Today, the television technician regards a dual-channel triggered-sweep oscilloscope as a basic service instrument. Bread-and-butter types of service oscilloscopes have vertical-amplifier bandwidths of approximately 5 MHz. Oscilloscopes designed for servicing computer-type circuitry, such as the one shown in Fig. I-2, provide vertical bandwidth in the order of 30 MHz. Professional varieties of oscilloscopes designed for compre-

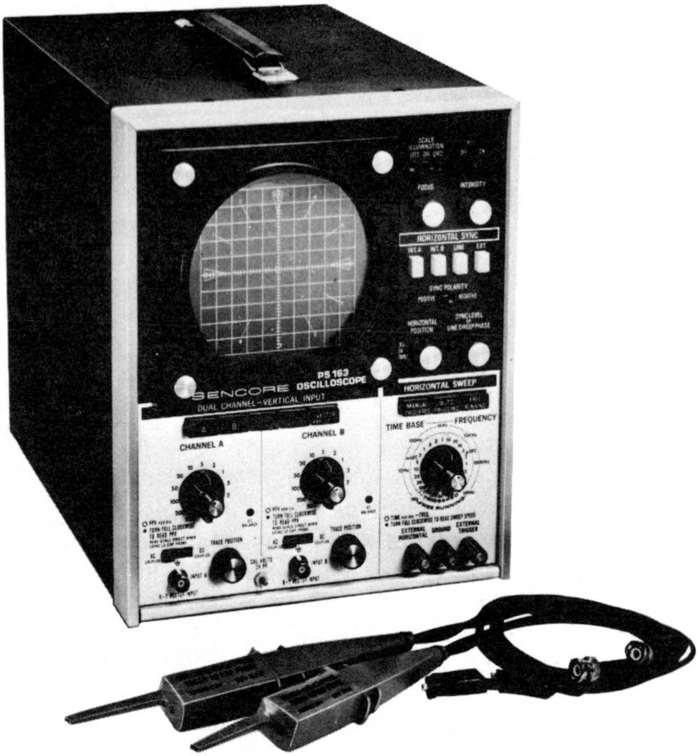

Courtesy Sencore, Inc.

Fig. I-1. A modern service-type oscilloscope.

hensive testing of high-speed digital circuitry (Fig. I-3) feature vertical-amplifier bandwidths in the order of 100 MHz. "Old old timers" are necessarily impressed also by the comparatively small dimensions and light weight of extra-high performance modern oscilloscopes, with respect to the large and heavy low-performance scopes used forty years ago.

This book emphasizes practice, not theory. It is designed to show you how to do various jobs with your equipment as efficiently as possible. Information is presented without frills or double talk. Since a television set has the most involved electronic circuitry the average technician encounters, much of the text is related to testing the various sections and components of this type of receiver. This book has a twofold purpose: to help you understand how to make waveform tests with an oscilloscope, and to show you how to analyze the waveforms produced by defective circuits. A careful study of these pages

Courtesy B&K Precision Products of Dynascan Corp.

Fig. I-2. A 30-MHz oscilloscope suitable for digital-circuitry tests.

should make your work easier and more effective. Beginners will find it very helpful to make the various tests at the bench, as they proceed from one topic to the next. This provides a form of reinforced learning which facilitates both learning and retention of new material.

Most servicing is done with voltmeters. *An oscilloscope is a voltmeter. It is a more complete voltmeter than a vom or dvm.* A scope gives more complete information than a vom (or dvm), because it shows how a voltage rises and falls in a receiver circuit. This variation is called the waveform of the voltage.

It is sometimes supposed that a scope is difficult to operate. Admittedly, a modern scope with triggered sweeps and dual-channel display is more elaborate than the obsolescent instruments with free-running horizontal-deflection oscillators and single-channel display. However, it is still easier to learn how to use a scope than to learn how to ride a bicycle.

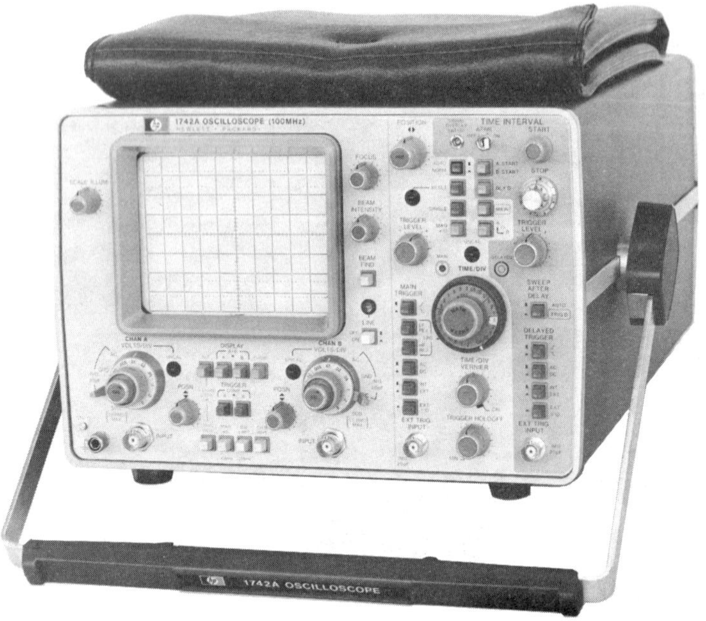

Courtesy Hewlett-Packard

Fig. I-3. A professional type of oscilloscope with a 100-MHz bandwidth.

Just as we can go places faster on a bicycle than on foot, we can troubleshoot tv receivers faster with a scope than with a voltmeter. For example, a meter cannot show whether a voltage is "ringing"—a scope does. A meter cannot show whether a signal is undistorted, clipped, or noisy—but a scope does. A meter cannot indicate the occurrence of parasitics, cross talk, or phase shift—a scope shows these troubles at a glance.

Although this book is not intended to be used as a textbook, if you have never used a scope before you will find it helpful to read the next few pages carefully. Here you will find a practical discussion of peak-to-peak, instantaneous, effective, and average values of sine waves, and how they tie in with the peak-to-peak values of complex tv waveforms.

Television service literature frequently refers to "peak-to-peak" voltage values. The ac voltage scales of a vom indicate *rms* voltage values. The meter movement in a vom responds to the *average* value of a rectified sine wave. Similarly, a rectified sine wave is displayed on the screen scope with its average value at the zero (beam-resting) level, when the scope is operated on its ac function. What do these various terms mean?

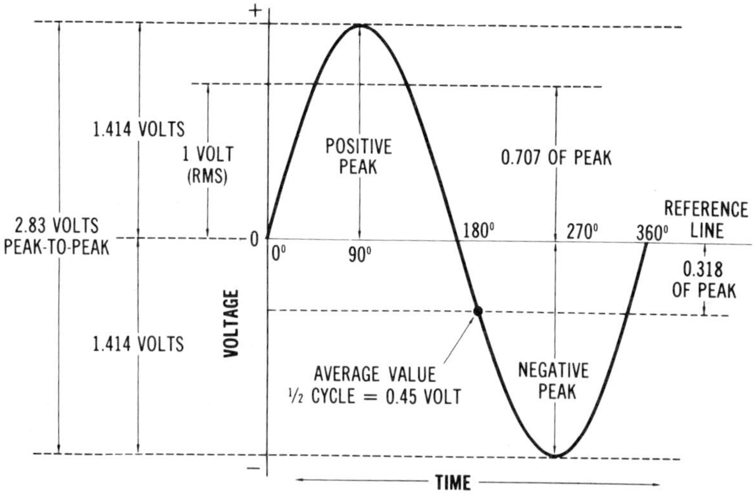

Fig. I-4. A sine wave with important values.

A sine wave, with its important component values indicated, is given in Fig. I-4. The highest value reached in any one direction by a waveform is called its *peak* value. The *peak-to-peak* voltage is equal to the total excursion from the positive peak to the negative peak. The two peaks of the sine wave have the same values on both positive and negative half-cycles—that is, a sine wave is a *symmetrical* waveform. The peak voltage is equal to one-half of the peak-to-peak voltage. The rms voltage (indicated by a vom) is equal to 0.707 of the peak voltage, as will be explained later in greater detail.

Of course, not all television waveforms are symmetrical. For example, consider the basic pulse waveform as shown in Fig. I-5. Note that here the positive-peak voltage is not equal to the negative-peak voltage. However, the positive area of the waveform is equal to its negative area. The sum of the positive-peak and negative-peak voltages is equal to the peak-to-peak volt-

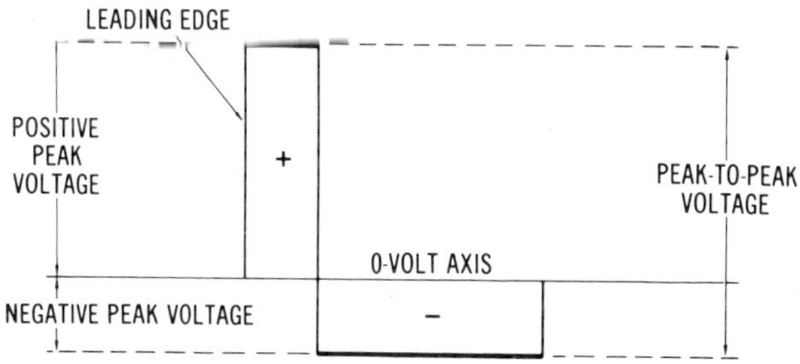

Fig. I-5. Basic pulse waveform.

age. Note that any complex waveform will be displayed on the scope screen with its positive-peak voltage above the zero-volt axis, and with its negative-peak voltage below the zero-volt axis, when the scope is operated on its ac function.

As can be seen in Fig. I-4, the *average value of a sine wave is zero*—that is, the average value over one complete cycle is zero. It is sometimes puzzling to consider how we can speak of "10 milliamperes of alternating current," because the instantaneous value of the current is continuously changing, and the average value of a complete cycle is zero. This puzzle is solved by first recognizing that, although the average value of an ac current is zero when tested with a dc meter, it does have some finite value when tested with an ac meter. Furthermore, when an alternating current is used for heating a soldering gun, or for lighting a crt filament for example, *both* the positive and the negative half-cycles of the waveform effectively produce heat and/or light, and do *not* cancel out. In other words, cancellation occurs only when the ac waveform is applied to a dc utilization device, such as a dc voltmeter.

Early in the development of the electrical industry, dc was used exclusively for heating, lighting, and power. Later, ac was found to be more economical to distribute and use, and ac power systems gradually displaced dc power systems. Now, the industry faced a new problem. A unit of voltage and current measurement was needed, whereby "110 volts" of ac would produce the same amount of heat, light, or power as "110 volts" of dc. What unit of ac measurement provides this equivalence?

Equivalence is realized when ac is measured in terms of "effective" values. In other words, the *effective* value of an ac current corresponds to a dc *value.* A soldering gun will get just as hot when energized by 110 volts of dc as it will by 110 effective volts of ac. Usually, we speak of an effective value as an rms value. As noted in Fig. I-4, the rms value is equal to 0.707 the peak value. Vom's are calibrated to read ac voltage and current in terms of rms values. The initials "rms" stand for "root mean square." Thus, we can compile the useful relations of *sine-wave* voltages:

PEAK-TO-PEAK VOLTAGE = 2 × PEAK VOLTAGE
PEAK VOLTAGE = 1/2 × PEAK-TO-PEAK VOLTAGE
RMS VOLTAGE = 0.707 × PEAK VOLTAGE
PEAK VOLTAGE = 1.414 × RMS VOLTAGE
PEAK-TO-PEAK VOLTAGE = 2.83 × RMS VOLTAGE

Furthermore, if you are using a dc scope, you need to know that when you apply a 1.5-volt battery across the vertical input

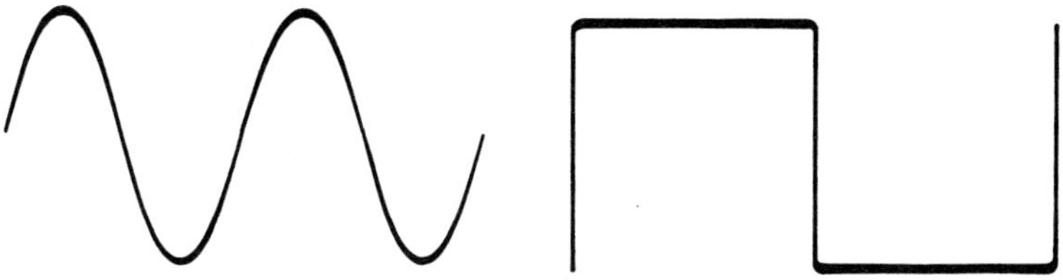

Fig. I-6. Sine wave and square wave with the same peak-to-peak voltage.

terminals of the dc scope, the trace will move the same vertical height as when a 1.5 peak-to-peak sine wave voltage is applied.

The peak-to-peak voltages of the sine wave and square wave in Fig. I-6 are equal. On the other hand, it is clear that the rms voltages of the two waveforms are *not* equal. We know that the rms voltage for the sine wave is equal to 0.707 of its peak voltage. It will be realized by inspection of the square wave that its rms voltage is equal to its peak voltage. As a general rule, the rms value of a complex waveform cannot be determined by inspection. Service procedures require that rms values be measured only when heat or mechanical power is to be produced. Otherwise, scope operators are concerned solely with peak and peak-to-peak voltages. Note, too, that an ac waveform may be associated with a dc level, as depicted in Fig. I-7. In this situation, a dc scope provides measurement of the dc component level and of the peak-to-peak voltage of the ac component (sine wave in this example).

As noted before, electronics technology is becoming increasingly more advanced. We must keep up with these advances if we are to remain competitive. Unless the full capabilities of oscilloscope use are clearly understood, it will become much more difficult in the future to properly service modern circuitry.

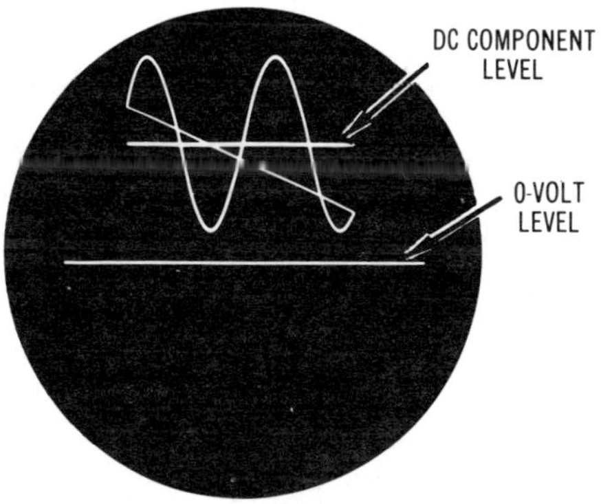

Fig. I-7. How an ac waveform with a dc component is displayed on the screen of a dc scope.

SECTION 1

Audio Tests and Measurements

1–1. To Check for Amplifier Distortion With a Dual-Trace Oscilloscope

Equipment: Audio Oscillator.

Connections Required: Connect audio oscillator output cable to input terminals of amplifier. Connect power resistor of suitable value across amplifier output terminals. Connect oscilloscope "A" channel across amplifier input terminals, and connect oscilloscope "B" channel across amplifier output terminals, as shown in Fig. 1-1.

Procedure: Advance output level from audio oscillator to obtain maximum rated power output from the amplifier. A basic distortion test is made at a frequency of 1 kHz. Adjust channel A and B gain controls to obtain patterns of equal height. Adjust the positioning controls to superimpose the two waveforms.

Evaluation of Results: If the two waveforms superimpose perfectly, the amplifier is practically distortionless. On the other hand, when there is a discrepancy between the two patterns, as shown in the diagram, the amplifier is distorting. *If the output pattern "writhes" when the audio oscillator is set to 59 Hz or 119 Hz, hum voltage is present.*

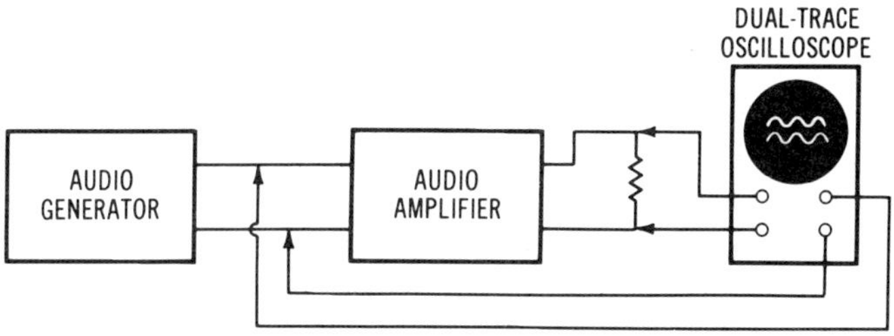

(*A*) *Test connections.*

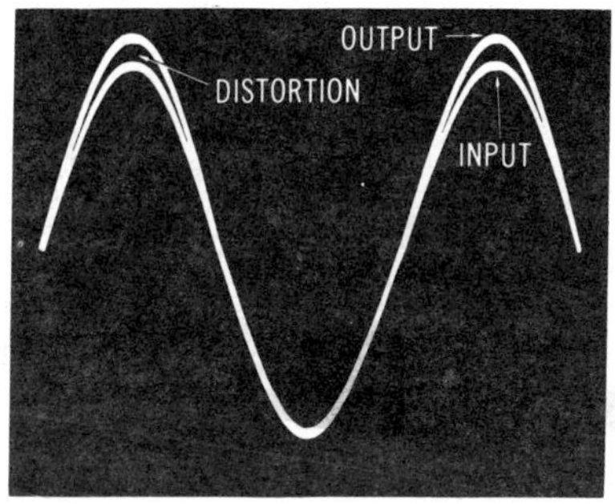

(*B*) *Superimposed input/output patterns.*

Courtesy Sencore, Inc.

Fig. 1-1. Checking for audio distortion with a dual-trace oscilloscope.

NOTE 1–1

Basic Controls of Dual-Trace Oscilloscope

A dual-trace oscilloscope uses triggered sweep, with the basic controls shown in Fig. 1-2. Some dual-trace scopes have separate time bases for the vertical channels. The foregoing distortion test must be made with both traces swept at the same speed. This distortion test can be made, if desired, with a conventional oscilloscope supplemented by an electronic switch.

1–2. To Check for Amplifier Distortion With a Differential-Input Scope

Equipment: Audio oscillator.

Connections Required: Same as in Fig. 1-1.

Procedure: Drive the amplifier to maximum rated power output at 1 kHz. Adjust channel A and B gain controls to obtain patterns with the same height. Then switch the oscilloscope for A-B display.

Evaluation of Results: If a straight horizontal line is displayed on the scope screen, the amplifier is practically distortion-

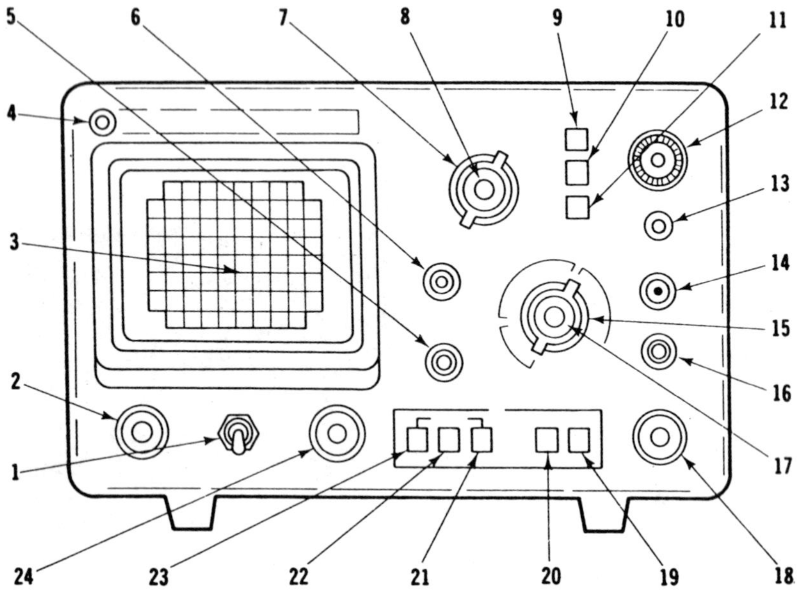

1. POWER ON-OFF toggle switch. Applies power to oscilloscope.

2. INTENSITY control. Adjusts brightness of trace.

3. Graticule. Provides calibration marks for voltage and time measurements.

4. Pilot lamp. Lights when power is applied to oscilloscope.

5. ◂▸ POSITION control. Rotation adjusts horizontal position of trace. Push-pull switch selects 5X magnification when pulled out; normal when pushed in.

6. ⬍ POSITION control. Rotation 3adjusts vertical position of trace.

7. VOLTS/DIV switch. Vertical attenuator. Coarse adjustment of vertical sensitivity. Vertical sensitivity is calibrated in 11 steps from 0.01 to 20 volts per division when VARIABLE 8 is set to the CAL position.

8. VARIABLE control. Vertical attenuator adjustment. Fine control of vertical sensitivity. In the extreme clockwise (CAL) position, the vertical attenuator is calibrated

9. AC vertical input selector switch. When this button is pushed in the dc component of the input signal is eliminated.

Fig. 1-2. Basic controls of a

10. GND vertical input selector switch. When this button is pushed in the input signal path is opened and the vertical amplifier input is grounded. This provides a zero-signal base line, the position of which can be used as a reference when performing dc measurements.

11. DC vertical input selector switch. When this button is pushed in the ac and dc components of the input signal are applied to vertical amplifier.

12. V INPUT jack. Vertical input.

13. ⏚ terminal. Chassis ground.

14. CAL ⎍ jack. Provides calibrated 0.8 V p-p square wave output at the line frequency for calibration of the vertical amplifier.

15. SWEEP TIME/DIV switch. Horizontal coarse sweep time selector. Selects calibrated sweep times of 0.5 μ SEC/DIV to 0.5 SEC/DIV in 19 steps when VAR/HOR GAIN control 17 is set to CAL. Selects proper sweep time for television composite video waveforms in TVH (television horizontal) and TVV (television vertical) positions. Disables internal sweep generator and displays external horizontal input in EXT position.

16. EXT SYNC/HOR jack. Input terminal for external sync or external horizontal input.

17. VAR/HOR GAIN control. Fine sweep time adjustment (horizontal gain adjustment when SWEEP TIME/DIV switch 15 is in EXT position). In the extreme clockwise position (CAL) the sweep time is calibrated.

18. TRIG LEVEL control. Sync level adjustment determines point on waveform slope where sweep starts. In fully counterclockwise (AUTO) position, sweep is automatically synchronized to the average level of the waveform.

19. TRIGGERING SLOPE switch. Selects sync polarity (+), button pushed in, or (−), button out.

20. TRIGGERING SOURCE switch. When the button is pushed in, INT, the waveform being observed is used as the sync trigger. When the button is out, EXT, the signal applied to the EXT SYNC/HOR jack 16 is used as the sync trigger.

21. TVV SYNC switch. When button is pushed in the scope syncs on the vertical component of composite video.

22. TVH SYNC switch. When button is pushed in the scope syncs on the horizontal component of composite video.

23. NOR SYNC switch. When button is pushed in the scope syncs on a portion of the input waveform. Normal mode of operation.

24. FOCUS control. Adjusts sharpness of trace.

Courtesy B&K-Precision Products of Dynascan

dual-trace oscilloscope.

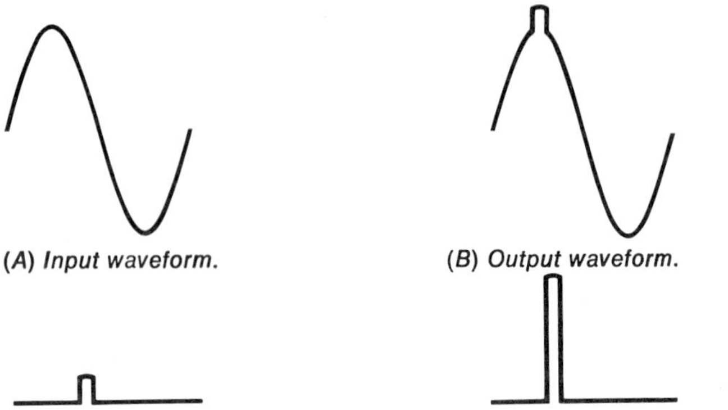

(*A*) *Input waveform.*

(*B*) *Output waveform.*

(*C*) *A—B waveform.*

(*D*) *Vertically expanded A—B waveform.*

Fig. 1-3. Difference waveform consists of distortion products.

less. On the other hand, if any waveform is displayed, the amplifier is distorting, as exemplified in Fig. 1-3.

NOTE 1–2

Vertical Expansion of Distortion Pattern

Small values of distortion are difficult to "see" unless the distortion waveform is expanded vertically on the scope screen. To do so, supplement the foregoing procedure by advancing the gain of both channels to maximum. Then, adjust the vernier gain control of one channel to

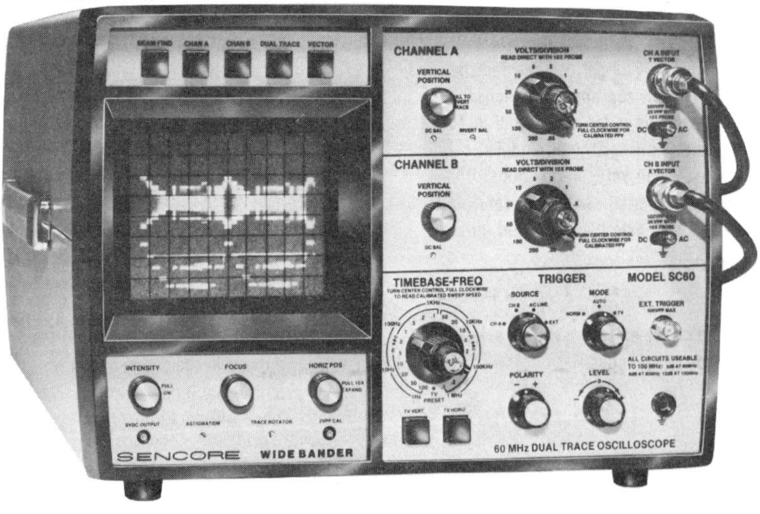

Fig. 1-4. A modern oscilloscope that provides dual-channel displays and a choice of A−B or A+B patterns.

minimize the amplitude of the expanded pattern. Any residual waveform that cannot be cancelled by adjustment of the vernier gain control is a distortion waveform.

NOTE 1–3

Input/Output Phase Relations

In the foregoing example, the amplifier input and output waveforms have the same phase reference. However, an amplifier is equally likely to have an output waveform that is 180° out of phase with the input waveform. In such a case, check the distortion products by switching the oscilloscope for A+B display. In other words, when −B is added to +A, the result is the same as subtracting +B from +A.

1–3. To Check for Amplifier Distortion With Lissajous Patterns

Equipment: Audio oscillator.

Connections Required: Connect output from audio oscillator to the input of the amplifier and to the horizontal-input terminals of the oscilloscope. Connect a suitable power-type resistor across the amplifier output terminals and to the vertical-input terminals of the oscilloscope, as shown in Fig. 1-5.

Procedure: Drive the audio amplifier to maximum rated output at 1 kHz, and adjust scope controls to display a diagonal line on the screen.

Evaluation of Results: If the displayed diagonal line is precisely straight, the amplifier is practically distortionless. On the other hand, a curved line indicates the presence of amplitude distortion.

NOTE 1–4

Amplifier Test Frequencies

It is standard practice to make audio-amplifier tests at a frequency of 1 kHz. However, more complete data can be obtained by repeating a test at a low frequency such as 20 Hz, and at a high frequency such as 20 kHz. To measure the power output from an amplifier, use the formula:

$$\text{Watts} = \frac{E^2}{R}$$

where,

E is the rms output voltage,
R is the ohmic value of the resistor connected across the amplifier output terminals.

The rms output voltage is equal to the peak-to-peak output voltage multiplied by 0.354 (scopes are generally calibrated to read peak-to-peak values). *Note that the oscilloscope amplifiers should have better characteristics than the amplifier under test. On the other hand, the audio oscil-*

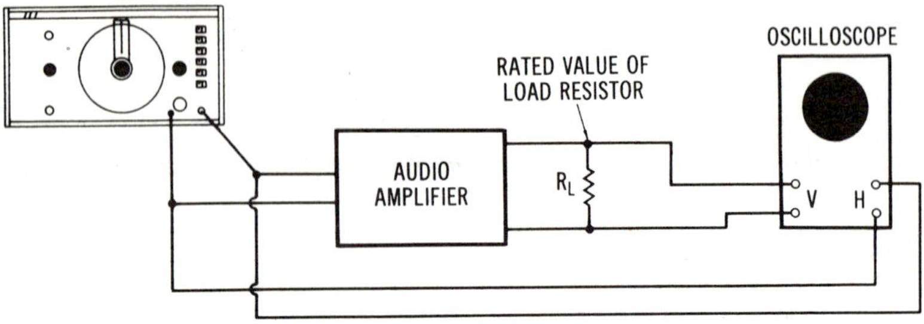

(A) Test setup.

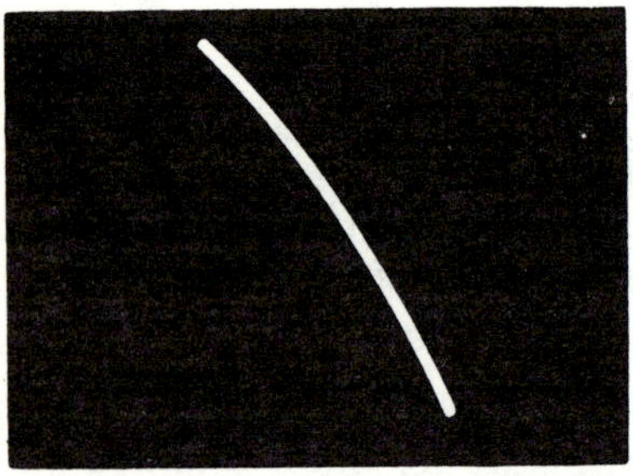

(B) Curved Lissajous line indicates amplifier nonlinearity.

Fig. 1-5. Distortion test.

lator can have more distortion than the amplifier under test, and the Lissajous pattern will still be precisely straight if the amplifier is distortionless.

1–4. To Check Amplifier Phase Shift With Lissajous Patterns

Equipment: Audio oscillator.

Connections Required: Connect output from audio oscillator to input of amplifier and to horizontal-input channel of oscilloscope. Connect output of amplifier to a suitable load resistor and to the vertical-input terminals of the scope, as shown in Fig. 1-6.

Procedure: Operate generator near the low-frequency cutoff point of the amplifier; adjust scope controls to obtain a centered Lissajous figure, as exemplified in the diagram. Then tune generator near the high-frequency cutoff point of the amplifier, and observe the resulting Lissajous figure.

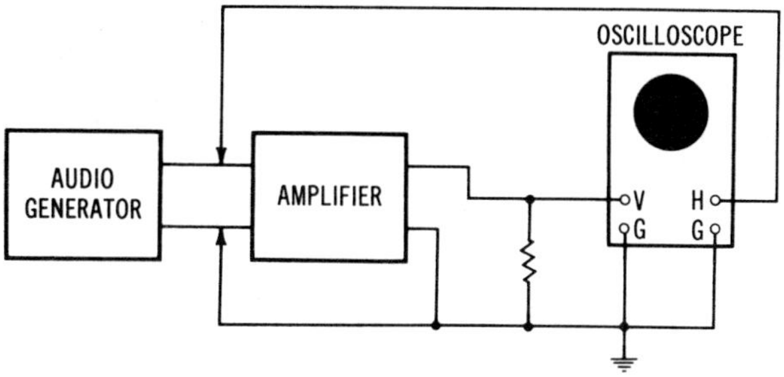

(*A*) *Test setup.*

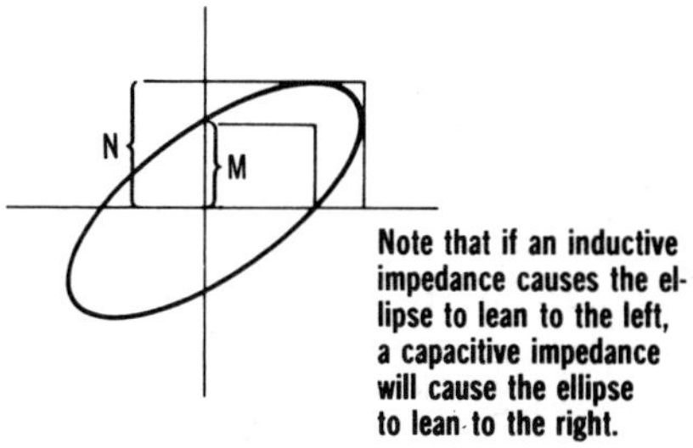

(*B*) *Phase-angle shift is equal to arcsin M/N.*

Fig. 1-6. Phase measurement of amplifier input/output relation.

Evaluation of Results: The phase-shift angle is equal to the angle whose sine (arcsin) has the value M/N, where M and N are the pattern excursions indicated in the diagram.

NOTE 1–5

Phase Shift Versus Test Frequency

An audio amplifier generally exhibits increasing phase shift from input to output as the test frequency approaches a cutoff value (−3 dB point). The amplifier will normally have zero phase shift at midband. If the amplifier is completely dc coupled, it will normally have zero phase shift at low frequencies, and will not exhibit a low-frequency cutoff point. Excessive phase shift at the band limits can lead to positive feedback via negative feedback loop(s), and thereby contribute to amplifier instability. When excessive phase shift occurs, look for defective capacitors; a faulty transistor can also introduce excessive phase shift. *Note that this method of phase measurement assumes that the pattern is a true ellipse (that both of the input waveforms to the scope are pure sine waves). If waveform distortion is present, the accuracy of phase measurement will be impaired.*

1–5. To Display Amplifier Distortion Products With a Conventional Scope

Equipment: Audio oscillator and harmonic distortion meter.

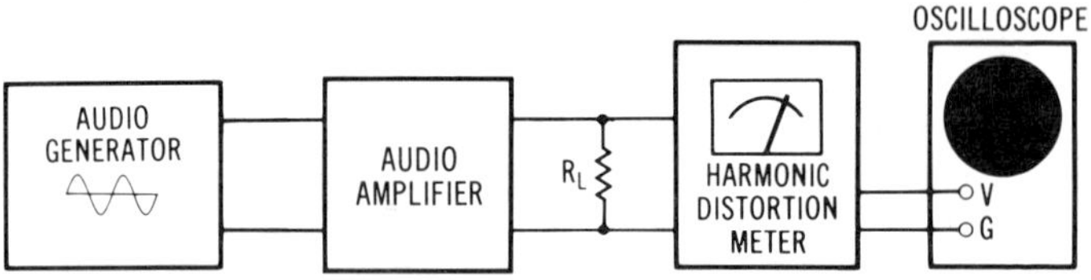

Fig. 1-7. Display of amplifier distortion products with a conventional scope.

Connections Required: Connect equipment as shown in Fig. 1-7.

Procedure: Tune audio oscillator for 1-kHz output, and drive amplifier to maximum rated power output. Tune harmonic distortion meter to reject the test frequency. Adjust scope controls to display the distortion products from the amplifier on the crt screen.

Evaluation of Results: The distortion pattern will show whether the distortion products consist chiefly of second-harmonic or third-harmonic, 60-Hz or 120-Hz hum, noise, or sum-and-difference voltages. In turn, the scope pattern provides helpful clues concerning the nature of the fault that is causing an abnormally high percentage of distortion. *Note that this test method requires that the audio oscillator output have less distortion than the amplifier under test. Otherwise, the distortion in the audio-oscillator waveform will be falsely charged to the amplifier under test.*

1–6. To Check an Amplifier for Music-Power Capability

Equipment: Pulse generator.

Connections Required: Connect output of pulse generator to input of amplifier. Connect a load resistor of suitable value across the amplifier output terminals and to the vertical-input channel of an oscilloscope, as shown in Fig. 1-8.

Procedure: Set the pulse generator for a pulse width of 1 millisecond and a repetition rate of 100 pulses per second. Advance the output from the generator until the top of the displayed pulse begins to show evidence of tilt. Then, back off slightly on the generator output amplitude.

Evaluation of Results: The music-power capability of the amplifier is equal to:

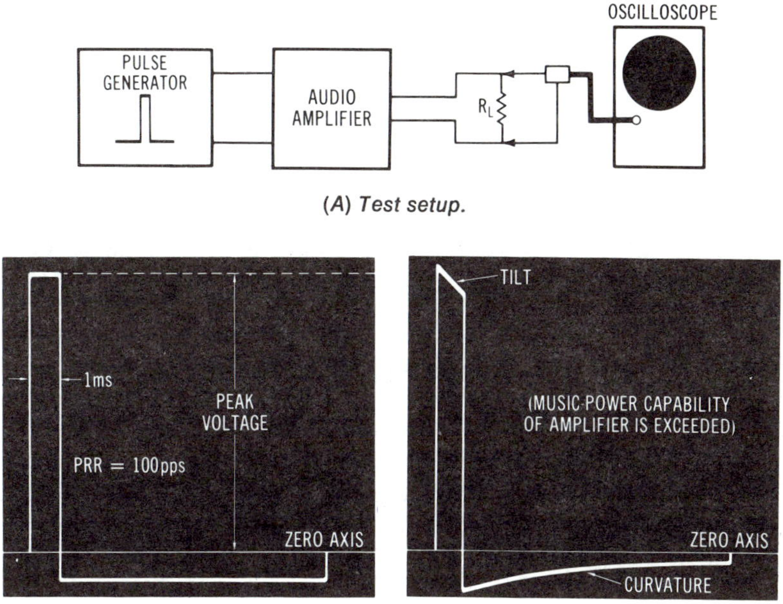

(*A*) *Test setup.*

(*B*) *Undistorted output pulse.*

(*C*) *Distorted output pulse.*

Fig. 1-8. Pulse test of hi-fi amplifier for determination of music-power capability.

$$\frac{E_p^2}{R_L}$$

where,

E_p is the peak voltage of the displayed pulse,
R_L is the value of the load resistor.

NOTE 1–6

Basis of Music-Power Capability

The music-power capability of an amplifier relates to its ability to reproduce sudden tonal surges or transient peaks that exceed the sine-wave (rms) capability of the amplifier's power output. The music-power capability of an amplifier is always greater than its rated power output value. The music-power value is primarily dependent upon the amount of dc power that is stored in the filter-output capacitors. In other words, a high-amplitude pulse demands a power surge that cannot be steadily maintained by the rectifier-filter system.

1–7. To Check the Square-Wave Response of an Amplifier

Equipment: Square-wave generator.

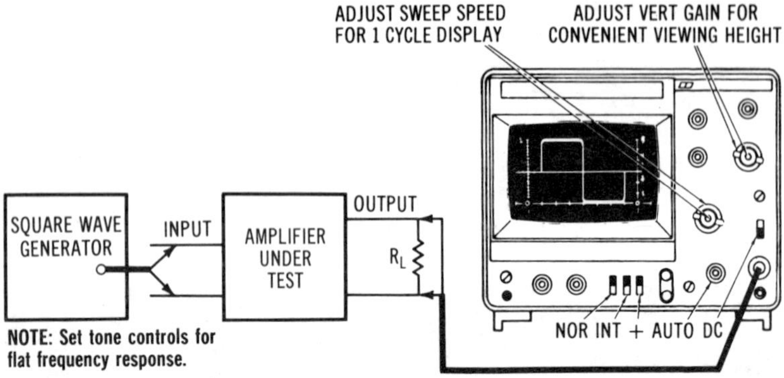

Fig. 1-9. Test setup for square-wave response.

Connections Required: Terminate amplifier in the rated load resistance and connect instruments as shown in Fig. 1-9.

Procedure: Drive amplifier to maximum rated power output at 1-kHz repetition rate. Note that the power output from the amplifier is equal to:

$$\frac{E_p^2}{R_L}$$

where,

E_p is the peak voltage of the square wave,
R_L is the value of the load resistance.

Observe the displayed waveform.

Evaluation of Results: An amplifier normally displays a virtually undistorted square-wave response at 1 kHz. Some manufacturers also rate their amplifiers for square-wave distortion at 20 Hz, and at 20 kHz. In typical designs, more or less tilt becomes evident at 20 Hz, and noticeable corner rounding appears at 20 kHz. Significant departure from rated square-wave response indicates a component or device defect in the amplifier network (Fig. 1-10).

NOTE 1–7

Significance of a Ringing Frequency

The ringing frequency in a reproduced square wave identifies the section or branch of an amplifier which has the highest Q value in the system, and which has L and C values that correspond to the ringing frequency. A circuit or section will ring if its associated resistance is sufficiently small that the configuration is underdamped. The ringing frequency in a reproduced square wave can be measured with a triggered-sweep scope. If the frequency-response curve of an amplifier has a hump or peak, a reproduced square wave will ring at the peak fre-

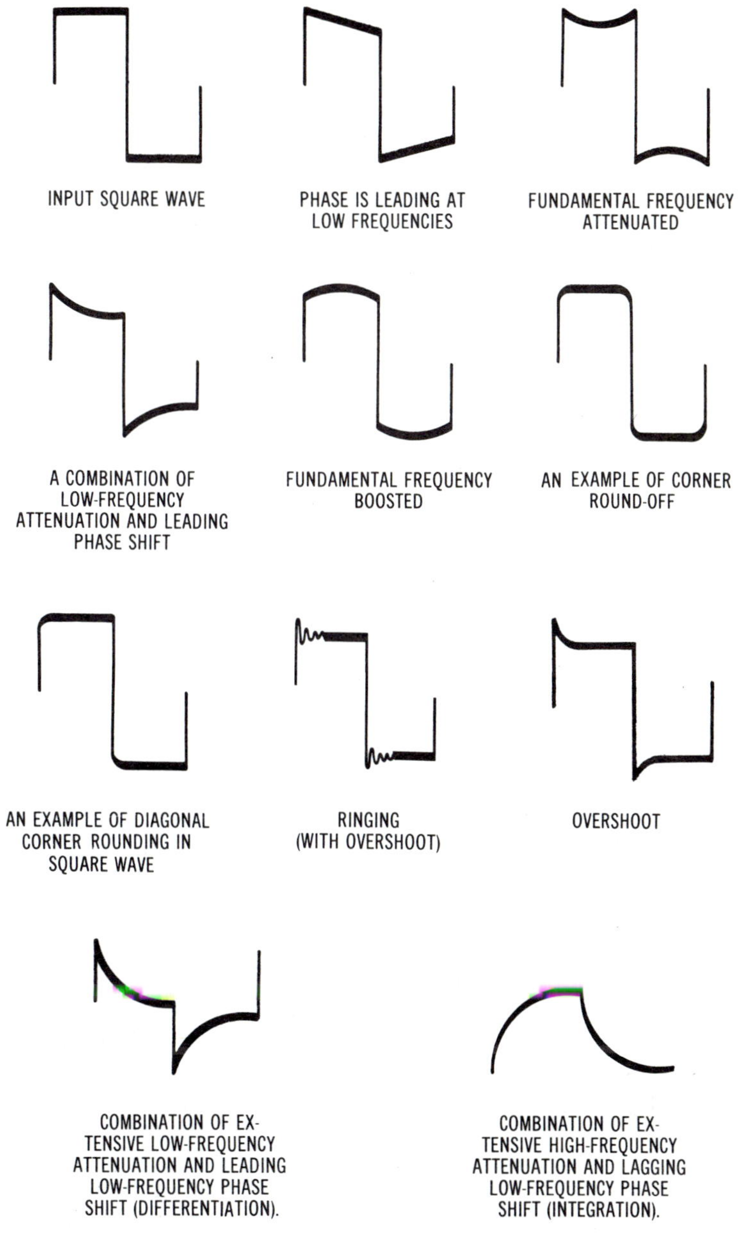

Fig. 1-10. Basic types of square-wave distortion.

quency. The extent of ringing depends on how pronounced the peak excursion may be.

1–8. To Check the Power Bandwidth of an Amplifier

Equipment: Audio oscillator and harmonic distortion meter.

Connections Required: Connect output from audio oscillator to input of amplifier under test. Terminate amplifier in its rated value of load resistance, and connect harmonic distortion meter across the load resistor. Connect oscilloscope at input of distortion meter, as shown in Fig. 1-11.

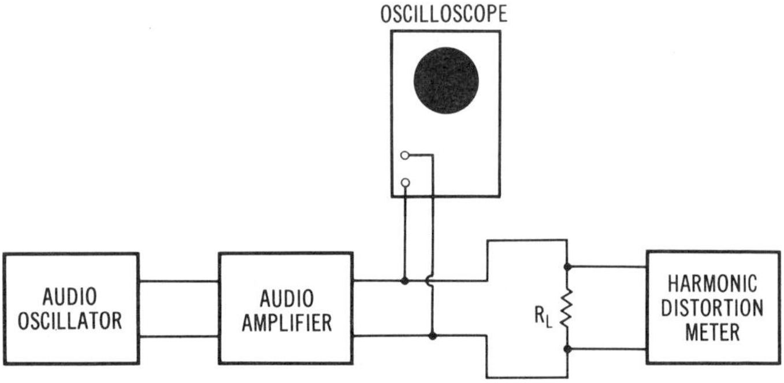

Fig. 1-11. Test setup for checking power bandwidth.

Procedure: Set audio oscillator to 1 kHz, and advance the input signal level to obtain maximum rated power output from the amplifier. Observe the percentage distortion reading indicated by the harmonic distortion meter. Then, reduce the output from the audio oscillator until the scope shows 84 percent of the output voltage for maximum rated power output. (This corresponds to three-quarters of maximum rated power output.) Vary the audio-oscillator frequency to determine the lower frequency and the upper frequency at which the amplifier now develops the same value of harmonic distortion as it did at 1 kHz when operating at maximum rated power output.

Evaluation of Results: The difference between the lower frequency and the upper frequency in the foregoing procedure is equal to the power bandwidth of the amplifier.

NOTE 1–8

Fidelity Versus Bandwidth

An amplifier generally has less distortion as its power output is decreased. At any output level, the distortion will increase as the test frequency approaches the high-frequency or the low-frequency cutoff point. This is just another way of saying that the high-fidelity bandwidth of an amplifier (power bandwidth) increases as the power output is decreased. As noted above, the standard power-bandwidth rating of an amplifier is specified at three-quarters of maximum rated power output. *Crossover distortion is basically different from overload distortion, in that when crossover distortion is present, the corresponding percentage of distortion will increase as the amplifier's power output is decreased.*

1–9. To Make a Stereo Separation Test

Equipment: Stereo signal generator.

Connections Required: Apply output from stereo generator to decoder (or to input of fm receiver). Connect the vertical-input lead from the oscilloscope to the L and R channel outputs, in turn.

Procedure: Apply R signal to decoder, and check the outputs from the L and R channels. Then apply an L signal to the decoder, and check the outputs from the L and R channels (Fig. 1-12).

Evaluation of Results: Ideally, there would be zero output from the L channel when an R signal is applied, and vice versa. In practice, separation is never perfect. A separation of 30 dB (a voltage ratio of approximately 32 times) is typical of high-quality stereo equipment.

1–10. To Measure the Output Resistance of an Amplifier

Equipment: Audio oscillator and two 8-ohm power-type resistors.

Connections Required: Connect the output from the audio oscillator to the input of the amplifier. Connect one 8-ohm resistor to the amplifier output terminals, and connect the scope across the load resistor (Fig. 1-13). (The other 8-ohm resistor will be shunted across the load in the second part of the test).

Procedure: Drive the amplifier to approximately half of its maximum rated power output at 1 kHz, and note the peak-to-peak

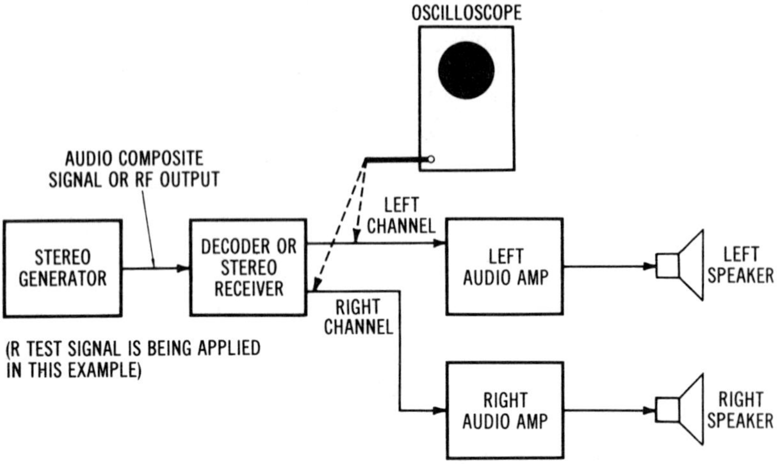

(*A*) *Equipment connections.*

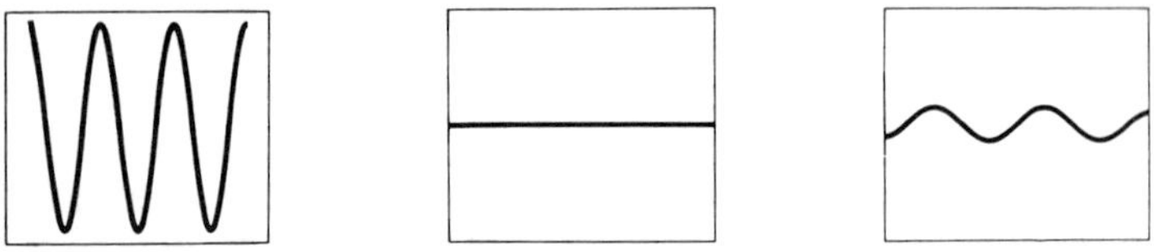

(*B*) *Display of R channel output.*

(*C*) *Ideal L channel output.*

(*D*) *Display of incomplete separation.*

Fig. 1-12. Stereo separation test.

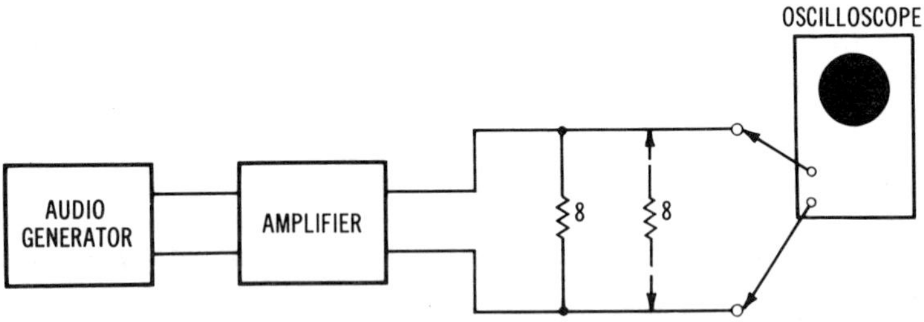

Fig. 1-13. Measurement of amplifier output resistance.

output voltage. Then connect the second 8-ohm resistor in parallel with the first resistor, and again note the peak-to-peak output voltage.

Evaluation of Results: Almost all audio amplifiers can work into an 8-ohm or a 4-ohm load. (A pair of 8-ohm resistors in parallel provides a 4-ohm load.) Multiply the foregoing peak-to-peak voltages by 0.354 to obtain the rms output voltages. Then, if the rms voltage across the 8-ohm load is E_1, and the rms voltage across the 4-ohm load is E_2, and R_o is the amplifier output resistance, calculate:

$$R_o = \frac{8(E_1 - E_2)}{2E_2 - E_1} \text{ ohms}$$

For example, if $E_1 = 6$ V rms, and $E_2 = 5$ V rms, then $R_o = 2$ ohms.

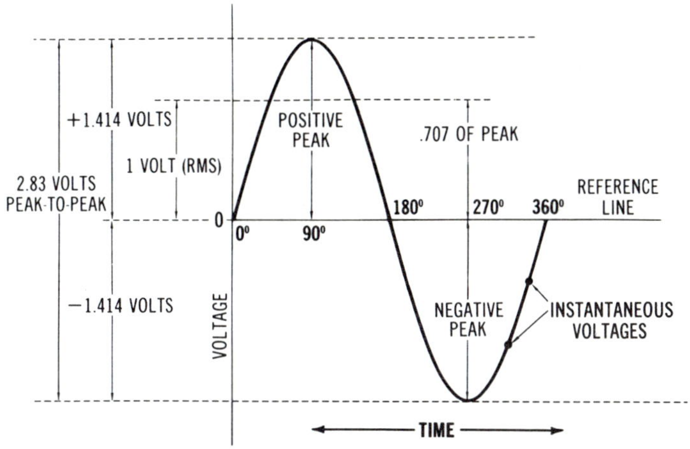

Fig. 1-14. Voltage values in a sine wave.

NOTE 1–9

Maximum Power Transfer

Maximum power is transferred from an amplifier to a load when the load resistance is equal to the amplifier's output resistance. On the other hand, high-fidelity speaker operation generally requires that the amplifier output resistance be 1/4 (or less) of the speaker impedance. The ratio of speaker impedance to amplifier output impedance (resistance) is called the damping factor of the system. Unless the speaker is adequately damped, its acoustic output will be distorted with respect to the audio output waveform from the amplifier. Although an ohm is based on rms voltage units, it is permissible to use peak-to-peak voltage values in calculation of R_o because the 0.707 conversion factor will automatically cancel out.

1–11. To Measure the Input Resistance of an Amplifier

Equipment: Audio oscillator and potentiometer with a range at least as high as the anticipated value of input resistance; load resistor; ohmmeter.

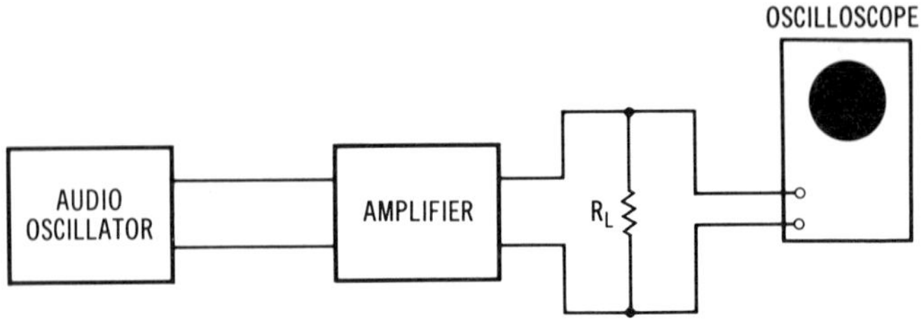

(A) Output voltage is measured at 1 kHz.

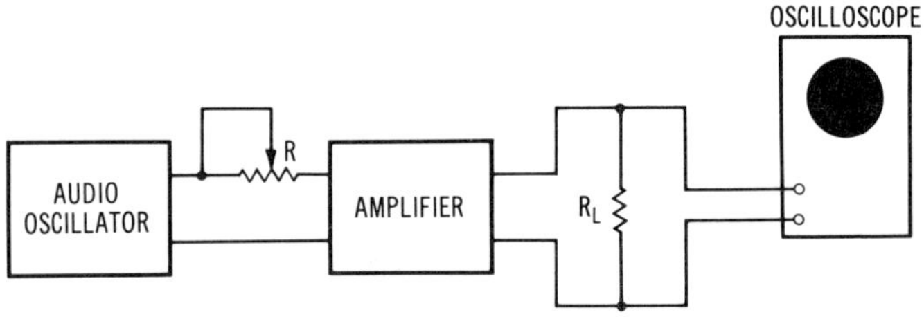

(B) Potentiometer resistance inserted to reduce output voltage to one-half.

Fig. 1-15. Measurement of amplifier input resistance.

Connections Required: With reference to Fig. 1-15, connect the output from the audio oscillator to the input of the amplifier. Connect the load resistor (typically rated at 10K) to the output of a preamplifier, or a power-type load resistor (typically rated at 8 ohms) to the output of a power amplifier. Connect scope across the load resistor. In the second part of the test, connect the potentiometer in series with the input to the amplifier.

Procedure: Advance the output from the audio oscillator to obtain rated maximum power output from amplifier at 1 kHz. Then, insert the potentiometer and adjust its resistance to obtain one-half the output voltage noted in the first part of the test.

Evaluation of Results: Measure the potentiometer resistance with an ohmmeter. The value of potentiometer resistance equals the input resistance of the amplifier.

NOTE 1–10

Audio Oscillator Output Resistance

The foregoing test procedure assumes that the output resistance of the audio oscillator can be disregarded with respect to the input resistance of the amplifier. In the majority of cases, this is permissible. However, for a precise measurement, the output resistance of the audio oscillator should be subtracted from the potentiometer resistance to calculate the amplifier input resistance. Thus:

$$R_{in} = R - R_s$$

where,

R_{in} is the amplifier input resistance,
R is the potentiometer resistance measured by the ohmmeter,
R_s is the source (output) resistance of the audio oscillator.

1–12. To Measure the Output Resistance of a Generator

Equipment: Potentiometer within the range of anticipated output resistance value; ohmmeter.

Connections Required: With reference to Fig. I-16, connect the output terminals of the generator to the input terminals of a scope. Then, connect the potentiometer across the output terminals of the generator. A 1-kHz test frequency is used.

Procedure: Note initial output-voltage reading. Then, note the reduced reading obtained when the potentiometer is connected to the output terminals of the generator. Adjust the potentiometer to obtain one-half the initial output-voltage reading.

Evaluation of Results: Remove the potentiometer and measure its resistance with an ohmmeter. The potentiometer resistance value is equal to the generator output resistance.

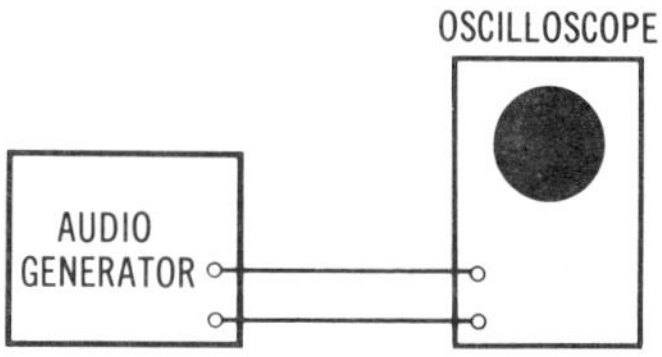

(A) Open-circuit output voltage is measured.

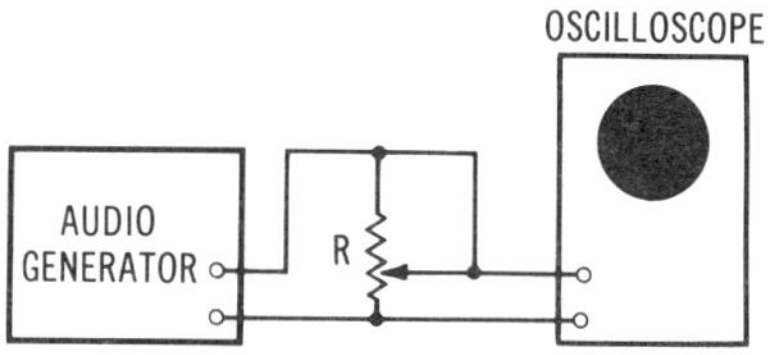

(B) Potentiometer is adjusted to reduce the initial reading to one-half.

Fig. 1-16. Measurement of generator output resistance.

1–13. To Check the Damping of a Speaker

Equipment: Square-wave generator and 75-ohm, 2-watt potentiometer.

Connections Required: Connect output from square-wave generator to potentiometer. Connect output from potentiometer to speaker and to the vertical-input channel of the scope (Fig. 1-17).

Procedure: Operate speaker in an "open-air location" or anechoic room to avoid interference from echoes. Observe the pattern on the scope screen at various square-wave repeti-

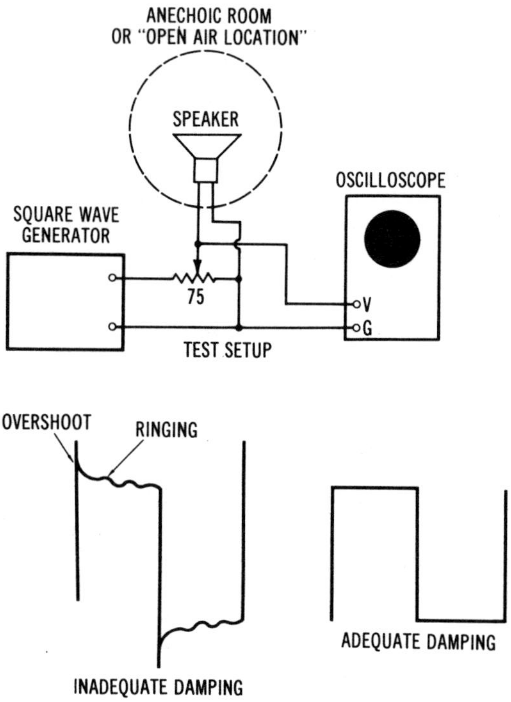

Fig. 1-17. Checking the optimum value of speaker damping resistance.

tion rates, and with comparatively low values of resistance across the speaker terminals.

Evaluation of Results: When the resistance shunting the speaker terminals is comparatively low, such as 1 or 2 ohms, overshoot and ringing will normally be minimized. In practice, the damping resistance is often chosen to provide a small amount of overshoot. This corresponds to the most rapid feasible rise (attack). Note that the optimum value of damping resistance determined in this test corresponds to the optimum value of output impedance for the amplifier with which the speaker is to be used.

NOTE 1–11

Speaker Sound Output Waveform

A speaker cone and cabinet arrangement acoustically couple the electrical energy in the voice coil to the surrounding air. The sound wave produced in the air by a square-wave voltage applied to the voice coil usually shows appreciable distortion. To check the sound waveform, operate the speaker in an anechoic room with a good-quality microphone, such as a condenser microphone with a preamplifier (as in a sound-level meter). Feed the output from the microphone preamplifier into a scope to display the acoustic output waveform from the speaker.

1–14. To Determine the Ringing Frequencies (Resonances) in a Listening Area

Equipment: Audio oscillator and good-quality microphone.

Connections Required: As depicted in Fig. 1-18, connect microphone to scope; connect audio oscillator to L speaker for test, then to R speaker. (Use microphone preamplifier, if necessary.)

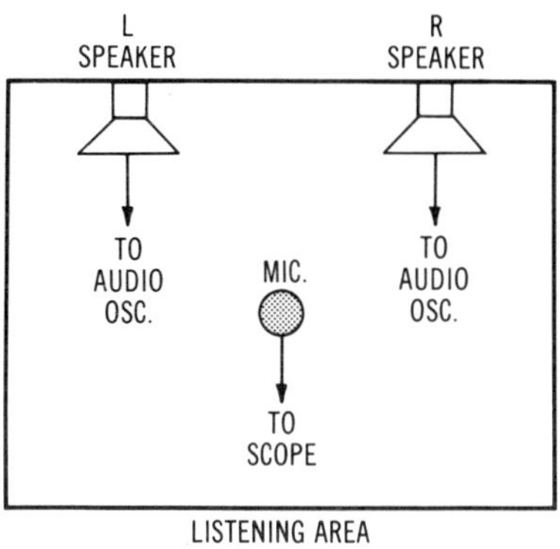

Fig. 1-18. Check of ringing frequencies (acoustic standing waves) in a listening area.

Procedure: With the microphone placed in the preferred listening position, vary the test frequency from the audio oscillator from 20 Hz to 20 kHz and observe the sound level from the L speaker on the scope screen. Then repeat the test for the R speaker.

Evaluation of Results: Every room (except an anechoic room) will exhibit standing waves (resonances) at various audio frequencies. In some types of rooms, these ringing frequencies are very prominent and impair the fidelity of sound reproduction. The frequency characteristic of a listening area can be improved by installation of drapes, curtains, and rugs in strategic areas. Speaker placement has a marked effect on standing-wave patterns; it is occasionally advantageous to locate the speakers in the corners of the room. *As the test frequency is varied and a room resonance is approached, the amplitude of the waveform displayed on the scope screen increases rapidly and then passes through a peak. The principal resonant frequency of a room corresponds to the frequency that produces the greatest pattern amplitude.*

SECTION 2

Impedance Measurements

2–1. To Measure the Impedance of a Speaker

Equipment: Audio oscillator, resistor.

Connections Required: With reference to Fig. 2-1, the speaker and resistor are connected in series across the push-pull output terminals of the audio oscillator. The voltage across the speaker is fed to the scope's vertical channel, and the voltage across the resistor is fed to the scope's horizontal channel.

Procedure: Use a suitable value of resistance (such as 8 ohms) to obtain a Lissajous figure of convenient size. Observe the value of the maximum vertical deflection and the value of the maximum horizontal deflection on the calibrated screen.

Evaluation of Results: The amount of vertical deflection indicates the peak-to-peak voltage applied across the speaker, and the amount of horizontal deflection indicates the peak-to-peak current flowing through the speaker. The impedance of the speaker is equal to the number of volts divided by the number of amperes. For example, if there are 2 volts dropped across the resistor, there is 1/4 ampere flowing; and if there are 4 volts dropped across the speaker, the impedance of the speaker is 16 ohms.

NOTE 2–1

Speaker Impedance Versus Frequency

The impedance of a speaker changes when the frequency of test is changed. Speaker impedance is customarily measured at 1 kHz. As the test frequency is increased, the impedance value ordinarily increases due to the fact that the voice coil is inductive, and inductive reactance increases with frequency. The phase angle between voltage and current may be measured as explained in 1–3. This phase angle also varies with changes in the test frequency. At one or more test frequencies, it will be

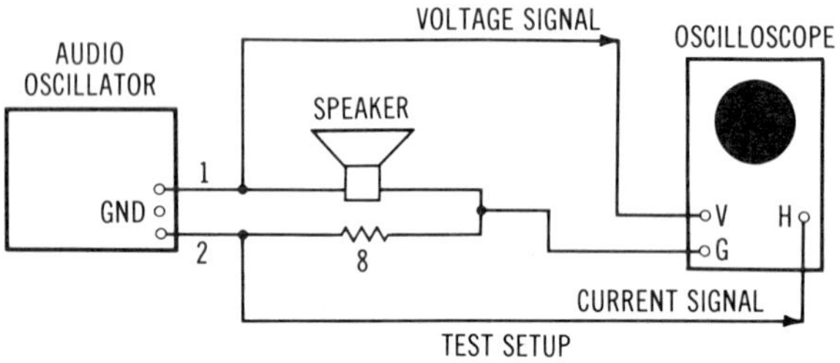

NOTE: Terminals (1) and (2) of the audio oscillator "float" with respect to ground. In other words, the Gnd terminal is <u>not</u> returned to any point of the (1)-(2) circuit in this type of instrument. The Gnd terminal is merely connected to the metal case of the audio oscillator.

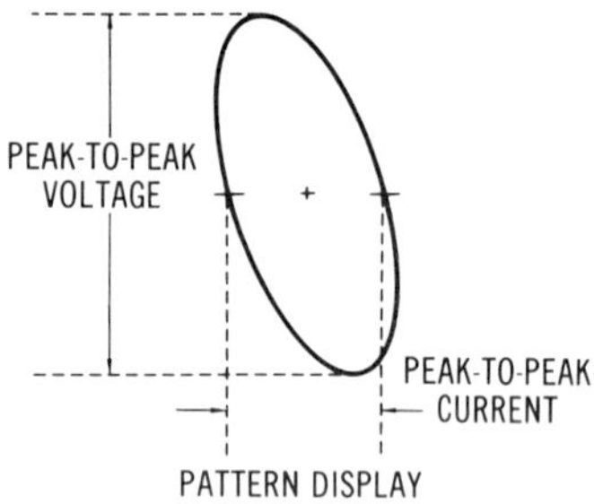

Fig. 2-1. Measurement of speaker impedance.

observed that the speaker is purely resistive (pattern is a straight diagonal line). When the speaker impedance becomes purely resistive, the reactance of the voice coil is being cancelled by the mechanical reactance of the vibrating cone assembly.

2–2. To Measure the Input Impedance of a Transformer

Equipment: Audio oscillator, resistor, normal load for transformer (such as a speaker).

Connections Required: Same as in Fig. 2-1. Audio oscillator drives the primary of the transformer under test, with the speaker connected to the secondary of the transformer. (See Fig. 2-2.)

Procedure: Same as in 2–1.

Evaluation of Results: Same as in 2–1.

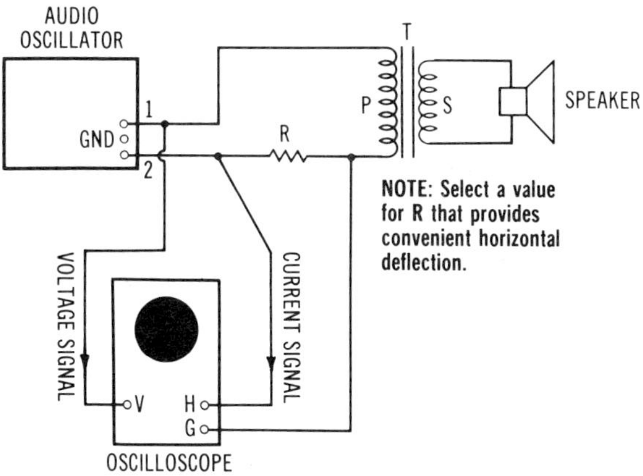

Fig. 2-2. Measurement of input impedance of a transformer.

NOTE 2–2

Transformer Turns Ratio, Voltage Ratio, Current Ratio, and Impedance Ratio

The turns ratio of a transformer is measured on open circuit (no load connected to the secondary). An ac voltage is applied to the primary; in turn, the ratio of primary voltage to secondary voltage is equal to the turns ratio of the transformer. Observe that the primary power (V_pI_p) is equal to the secondary power (V_sI_s). Thus, a step-down transformer with a 10-to-1 ratio, for example, will have a secondary voltage equal to 1/10 of the primary voltage, and will have a secondary current equal to 10 times the primary current. As the secondary current demand is increased, the primary current drain also increases. The primary will draw a very small current when the secondary is open-circuited; this small current demand results from the facts that the primary reactance is not infinite, that the primary winding has resistance (I^2R loss), and that the core of the transformer has a small power loss in its magnetic circuit. Note also that the primary/secondary impedance ratio of a transformer is equal to the square of its turns ratio. For example, a 10-to-1 transformer will have a primary/secondary impedance ratio of 100-to-1. Thus, if an 8-ohm speaker is connected to the secondary, the primary input impedance will be 800 ohms. Or, if a 4-ohm speaker is connected to the secondary, the primary input impedance will be 400 ohms.

2–3. To Measure the Output Impedance of a Transformer

Equipment: Audio oscillator and two resistors.

Connections Required: Connect the test setup as shown in Fig. 2-3.

Procedure: Same as in 2–2.

Evaluation of Results: Same as in 2–2.

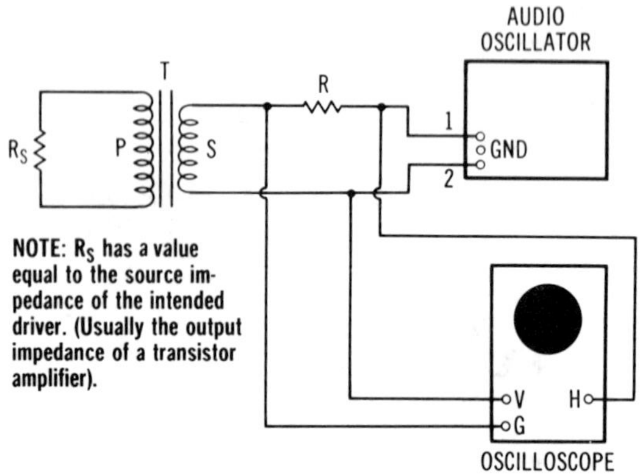

NOTE: R has any value that provides ample horizontal deflection on the scope screen.

Fig. 2-3. Measurement of output impedance of a transformer.

NOTE 2–3

Output Impedance and Maximum Power Transfer

The output impedance of a transformer for coupling an amplifier to a speaker is of importance because maximum power is transferred when the source impedance matches the load impedance. For example, suppose that the source (amplifier output) impedance is 1600 ohms, and that the speaker impedance is 16 ohms. Then, maximum power will be transferred to the speaker with a 10-to-1 output transformer. In other words, the impedance ratio of a 10-to-1 transformer is 100 to 1, so that a 1600-ohm source will be transformed into 16 ohms (which matches a 16-ohm speaker). In practice, output transformers are less than ideal—for example, an output transformer introduces phase shift in its low-frequency and high-frequency cutoff regions.

2–4. To Measure the Input Impedance Characteristic of a Crossover Network

Equipment: Audio oscillator, and resistor.

Connections Required: Refer to the diagram in Fig. 2-4; connections are the same as in 2–1 except that a tweeter speaker with a crossover capacitor C is included in the system.

Procedure: Vary audio-oscillator frequency through the range from 20 Hz to 20 kHz, and observe the extent of change in impedance pattern on the scope screen.

Evaluation of Results: When the value of the crossover capacitor C is optimum, a minimum change occurs in the imped-

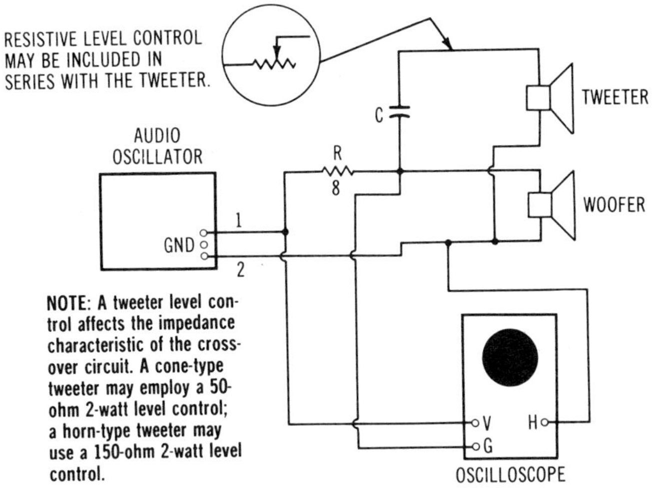

Fig. 2-4. Measurement of input impedance of a crossover network.

ance value as the test frequency is varied through the audio-frequency range.

NOTE 2–4

Crossover Circuit Operation

The simplest crossover circuit (which is sometimes adequate) is depicted in Fig. 2-4. A tweeter is connected in series with a crossover ca-

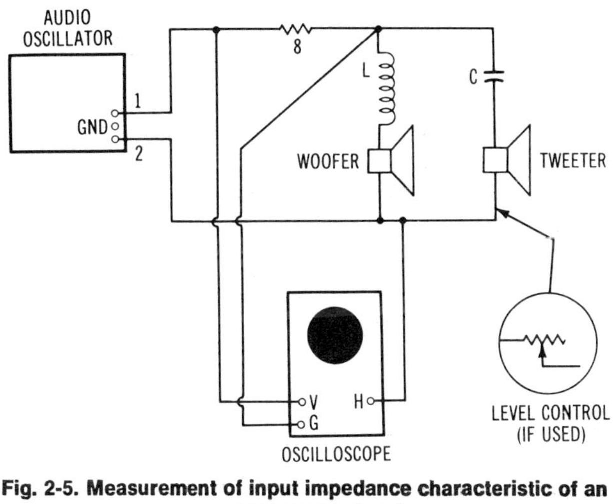

Fig. 2-5. Measurement of input impedance characteristic of an LC crossover network.

pacitor C, and the series arrangement is driven in parallel with the woofer. It is apparent that at some frequency, the reactance of the crossover capacitor will be equal to the tweeter impedance. This is called the crossover frequency. The crossover frequency depends on the value of the crossover capacitor. As an illustration, suppose that a 4-μF capacitor is used in series with an 8-ohm tweeter. Then, the crossover frequency will be 5 kHz, approximately. Or, if an 8-μF capacitor is used, the crossover frequency will be 2.5 kHz, approximately. At this crossover frequency, the tweeter consumes one-half of the power that it would consume with the crossover capacitor short-circuited. The crossover capacitor attenuates low audio frequencies, thereby preventing possible overload and damage to the tweeter. Although high audio frequencies enter the woofer, the overload potential is not as great. Note that better crossover network characteristics can be obtained by including an inductor in series with the woofer (see 2–5).

2–5. To Measure the Input Impedance Characteristic of an LC Crossover Network

Equipment: Same as in 2–4.

Connections Required: Connect the equipment to display an impedance pattern, as shown in Fig. 2-5.

Procedure: Same as in 2–4.

Evaluation of Results: When the values of the crossover capacitor C and the crossover inductor L are optimum, a minimum change occurs in the impedance value as the test frequency is varied through the audio-frequency range.

NOTE 2–5

Impedance Relations of an LC Crossover Network

An inductor has opposite reactance with respect to a capacitor. That is, as the frequency increases, inductive reactance increases. A crossover inductor may have a value of 0.25 mH; in turn, it will have an impedance of 8 ohms at 5 kHz (crossover frequency). Or, if the crossover inductor has a value of 0.5 mH, it will have an impedance of 8 ohms at 2.5 kHz (crossover frequency). Observe that when 8-ohm speakers are used in the arrangement of Fig. 2-5, the net impedance is not 4 ohms, because the crossover inductor L and the crossover capacitor C have different impedances at various frequencies. In normal operation, with correct values of L and C, the input impedance characteristic is practically constant at 8 ohms over the audio-frequency range. *At the crossover frequency, the inductor has an impedance of 8 ohms, and the capacitor has an impedance of 8 ohms; the network input impedance is 8 ohms. At 0.1 of the crossover frequency, the inductor has an impedance of 0.8 ohm, and the capacitor has an impedance of 80 ohms; the network input impedance is practically 8 ohms. At 10 times the crossover frequency, the inductor has an impedance of 80 ohms, and the capacitor has an impedance of 0.8 ohm; again, the network input impedance is practically 8 ohms.*

2–6. To Measure the Inductance of a Crossover Coil

Equipment: Same as in 2–1.

Connections Required: Same as in Fig. 2-1 (crossover coil is connected in place of the speaker).

Procedure: Use a value for the resistor that provides convenient horizontal deflection. Set the audio oscillator precisely to a suitable test frequency, such as 1 kHz. Observe the current value (resistive voltage drop), and the inductive voltage drop, as shown in the diagram of Fig. 2-6.

Evaluation of Results: The inductance of the coil is equal to the rms inductive voltage drop, divided by the rms current, and

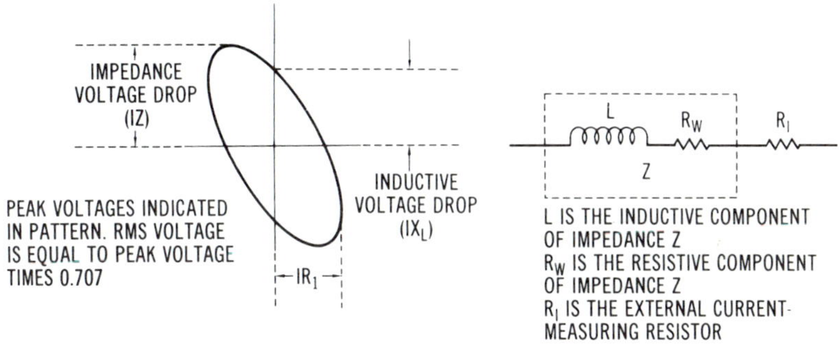

Fig. 2-6. Voltage indications in a Lissajous figure for a coil.

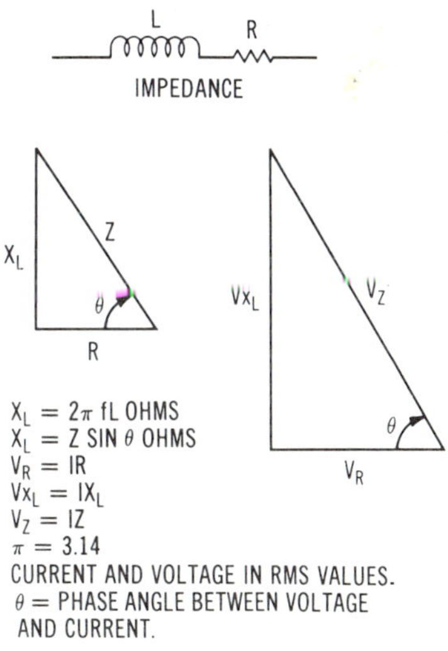

Fig. 2-7. Relations of resistance, inductance, and phase angle in coil in response to ac.

divided finally by $2\pi f$, where f is the test frequency. The rms voltage is equal to 0.707 times the peak voltage. Pi is equal to 3.14.

NOTE 2–6

Inductance, Resistance, Reactance, and Impedance

When working with inductive impedance (R and L in series), it is helpful to observe the relations of L, R, X_L, Z, V_R, V_{X_L}, and V_Z, as shown in Fig. 2-7. When an ac voltage is applied across R and L in series, the resistor drops IR volts, the inductor drops IX_L volts, and the impedance drops IZ volts. Thus, a coil that has an inductance of 0.5 mH will have an inductive reactance of 3.14 ohms at 1 kHz. Its impedance depends upon its winding resistance, as depicted in the diagram. Observe that the peak IX_L voltage drop is given in the Lissajous figure as shown in Fig. 2-6. The rms value of the IX_L voltage is equal to 0.707 times the peak IX_L voltage.

SECTION 3

Television Tests and Measurements

3–1. To Check the Video Signal for Distortion

Equipment: Test-pattern generator.

Connections Required: Connect output cable from generator to antenna-input terminals of tv receiver under test; connect output from video amplifier (or picture detector) to scope, as shown in Fig. 3-1.

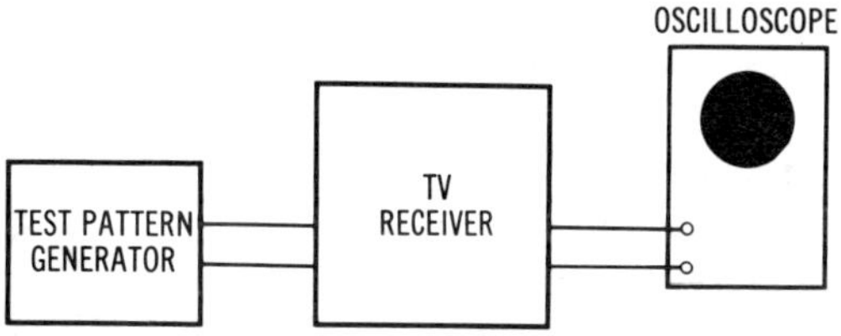

Fig. 3-1. Test setup for basic video signal distortion check.

Procedure: Display the video signal at field rate and at frame rate, as shown in Figs. 3-2 and 3-3.

Evaluation of Results: A standard test pattern contains peak whites; in turn, the horizontal sync tip should occupy 25 percent of the peak-to-peak signal excursion. If the sync tip occupies more than 25 percent of the signal amplitude, white compression and/or clipping is occurring. If the sync tip occupies less than 25 percent of the signal amplitude, sync

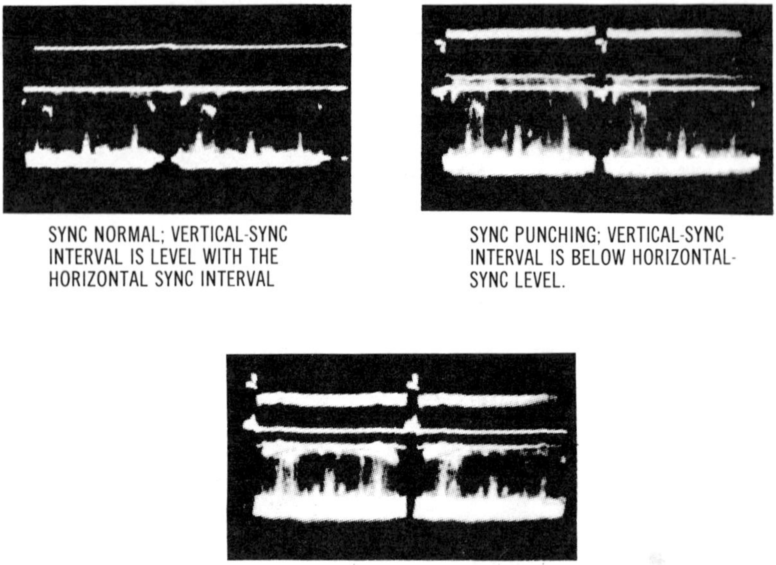

Fig. 3-2. Video signal at field rate.

compression and/or clipping is occurring. In a case where the vertical sync interval has less amplitude than the horizontal sync interval, sync punching is occurring. On the other hand, if the vertical sync interval has greater amplitude than the horizontal sync interval, horizontal sync attenuation is occurring. Incorrect percentage of horizontal sync tip with respect to the signal amplitude results from overload (often, incorrect bias voltage). Sync punching or its opposite distortion is caused by high-frequency attenuation or by low-frequency attenuation, respectively. Note that when the signal at the picture-detector output is normal, but is distorted at the video-amplifier output, the trouble will be found in the video amplifier.

3–2. To Check the Overall Signal Gain of a TV Receiver

Equipment: Calibrated am generator.

Connections Required: Connect output cable from generator to antenna-input terminals of tv receiver; connect output from video amplifier to oscilloscope, as shown in Fig. 3-4.

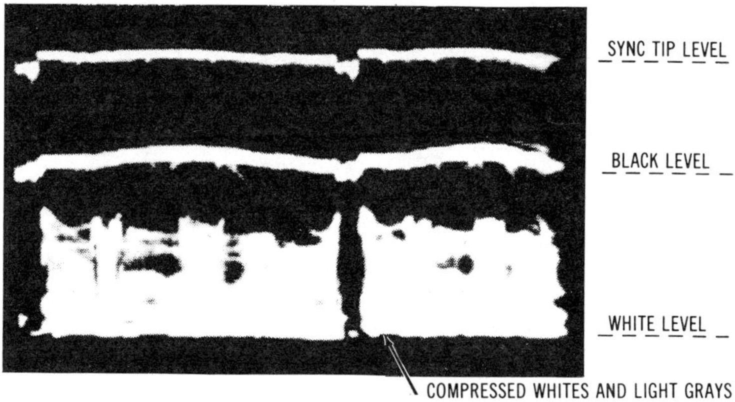

SYNC TIPS OCCUPY MORE THAN 25 PERCENT OF THE SIGNAL AMPLITUDE; WHITES AND LIGHT GRAYS ARE COMPRESSED (SYNC PUNCHING IS ALSO PRESENT)

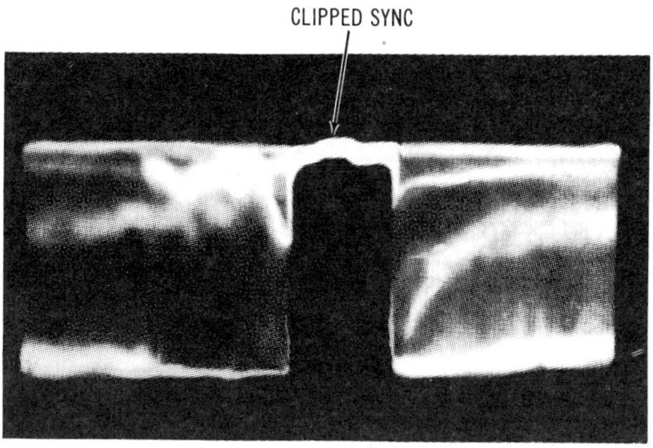

HORIZONTAL SYNC TIP IS PRACTICALLY ABSENT; EXAMPLE OF CLIPPED SYNC

Fig. 3-3. Video signal at field rate (above) and at frame rate (below).

Procedure. Tune generator to picture-carrier frequency of chosen channel, with 150 microvolts output and 90 percent modulation. Display the demodulated signal envelope on the scope screen (envelope frequency is usually 1 kHz).

Evaluation of Results: A typical black-and-white tv receiver operating at maximum available gain normally develops a 45-V p-p output in response to a signal input of 150 microvolts. Note that the output waveform should not have visible noise ("fuzz"). However, at lower input signal levels, noise will be-

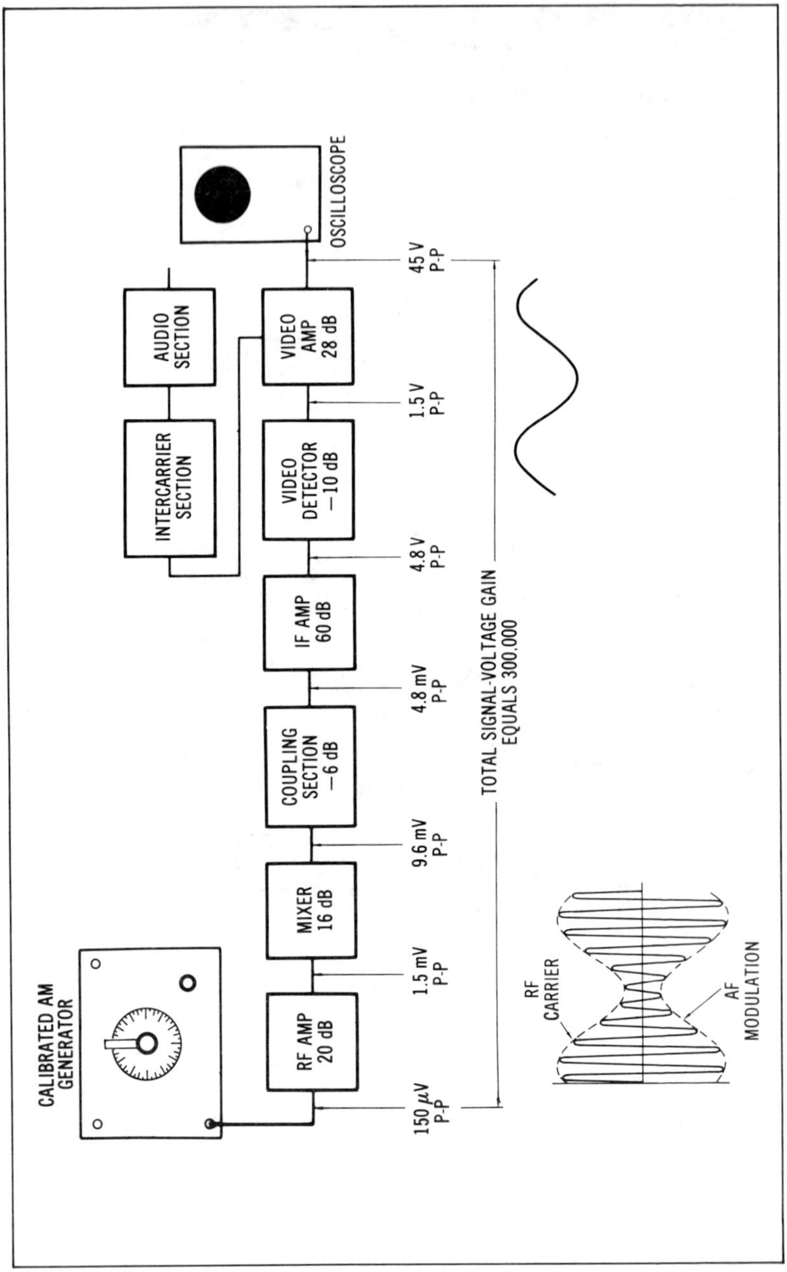

Fig. 3-4. Block diagram of typical black-and-white tv receiver operating at maximum available gain.

come evident on the output waveform. Excessive noise is often caused by collector junction leakage in a transistor.

3–3. To Measure the Sensitivity of a TV Receiver

Equipment: Calibrated signal generator.

Connections Required: Same as in Fig. 3-4 except that the oscilloscope is connected at the output of the picture detector.

Procedure: Observe the peak-to-peak voltage of the noise output from the picture detector when the generator signal is zero. Then, advance the output from the generator until the combined noise voltage and signal voltage is twice as great as previously noted for the noise voltage.

Evaluation of Results: From a practical standpoint, the useful sensitivity of a tv receiver is given by the number of microvolts signal input that is required to produce a picture-detector output (signal plus noise) which is double the value of the noise output in the absence of signal.

NOTE 3–1

The Four Fundamental Waveforms

The four fundamental waveforms are depicted in Fig. 3-5. The sine wave, square wave, and exponential wave are basically different from the noise waveform because they have a repetitive cyclic form, whereas a noise wave has a nonrepetitive random form. Sine, square, and exponential waveforms can be represented by simple algebraic equations, whereas a noise waveform must be described in terms of statistical functions. Although the peak-to-peak voltage of a sine wave, for example, can be precisely measured, the peak-to-peak voltage of a noise waveform can only be approximated because of its random characteristics.

3–4. To Signal-Trace the IF Section of a TV Receiver

Equipment: Demodulator probe; pattern generator will be helpful.

Connections Required: Connect output from pattern generator (or a tv antenna) to the receiver; connect demodulator probe to vertical-input terminals of oscilloscope.

Procedure: Set receiver controls for normal operation. Apply demodulator probe in turn at the collector of the first if stage, at collector of second if stage, and so on. Use high vertical gain and substantial generator output when checking the first if stage.

Evaluation of Results: When a demodulator probe such as depicted in Fig. 3-6 is used, and the scope is operated at a 30-Hz deflection rate, a "stripped" vertical-sync pulse will be

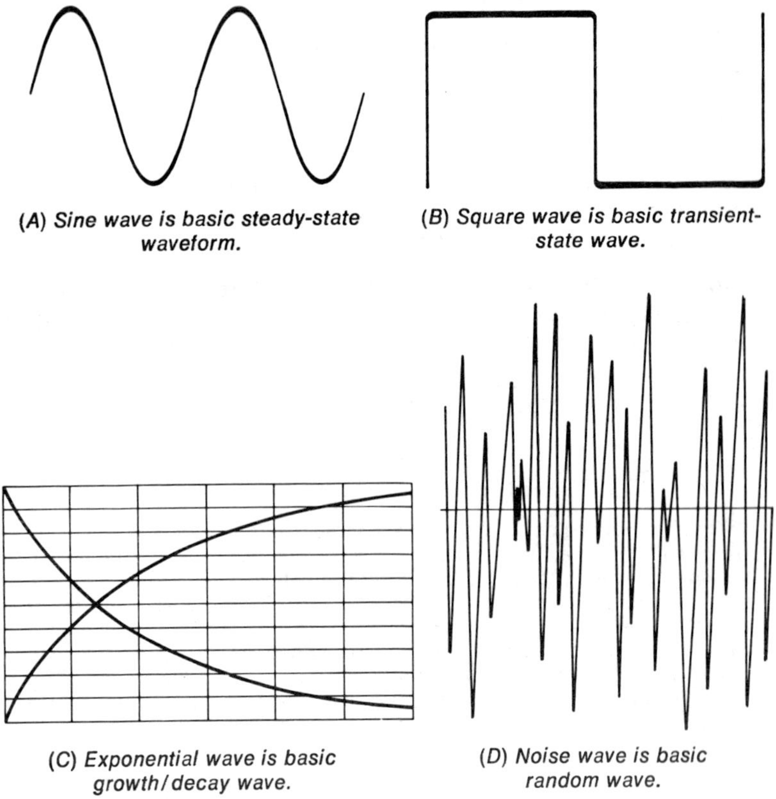

(A) Sine wave is basic steady-state waveform.

(B) Square wave is basic transient-state wave.

(C) Exponential wave is basic growth/decay wave.

(D) Noise wave is basic random wave.

Fig. 3-5. Four fundamental waveforms.

displayed on the scope screen, if the signal is present. Only a small residue of camera signal and horizontal sync pulses will be reproduced when this type of demodulator probe is used, because of its limited demodulating capability (Fig. 3-7). The chief advantage of this probe design is its comparatively high impedance. However, the probe has a marked loading and detuning action, and its application is limited to determining whether if signal is present or absent. Therefore, other methods must be used to measure stage gain.

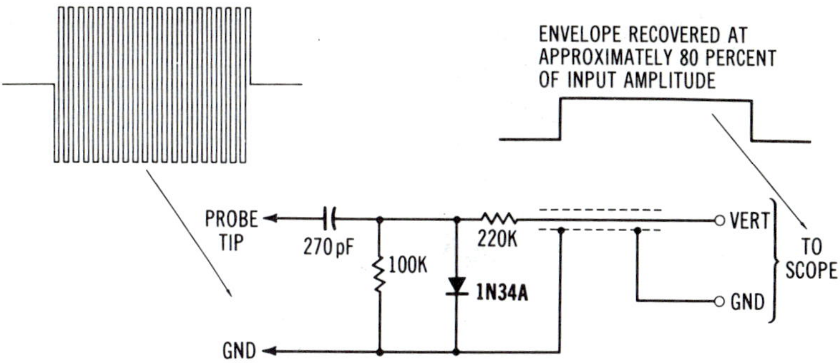

PROBE RESPONSE:
CARRIER FREQUENCIES FROM 500 kHz TO 200 MHz ENVELOPE (MODULATING) FREQUENCIES FROM 30 Hz TO 5 kHz PROBE INPUT CAPACITANCE: 2.25 pF, APPROX. PROBE INPUT RESISTANCE (APPROX.):

AT 500 kHz........................25 KILOHMS
1 MHz........................23 KILOHMS
5 MHz........................21 KILOHMS
10 MHz........................18 KILOHMS
50 MHz........................10 KILOHMS
100 MHz........................5.0 KILOHMS
150 MHz........................4.5 KILOHMS
200 MHz........................2.5 KILOHMS

MAXIMUM INPUT VOLTAGE: 28 PEAK VOLTS

WHEN USED IN VERY LOW LEVEL CIRCUIT TEST, A DEMODULATOR PROBE IS MORE EFFECTIVE IF A SMALL FORWARD BIAS (TYPICALLY 0.1 VOLT), IS APPLIED TO THE SEMICONDUCTOR DIODE.

Fig. 3-6. Standard demodulator probe configuration and characteristics.

NOTE 3–2

Simplest Type of Demodulator Probe

The simplest demodulator probe consists of a single diode, as shown in Fig. 3-8. However, it is not entirely practical because its demodulating capability is poor (the probe distorts a 60-Hz square-wave envelope substantially). Moreover, its application is limited to frequencies below 10 MHz, due to its development of standing waves on the oscilloscope input cable. Standing waves result in serious attenuation of the signal at various frequencies for which the cable is parallel-resonant.

3–5. To Monitor the IF Section of an Intermittent Receiver

Equipment: Demodulator probe(s) and/or lo-C probe.

Connections Required: With reference to the diagram in Fig. 3-9, connect a lo-C probe and a demodulator probe to the input terminals of a dual-trace oscilloscope. Connect the demodulator probe at the input of the third if stage, and connect the lo-C probe at the output of the picture detector. If two dual-trace scopes are available, connect demodulator probes to

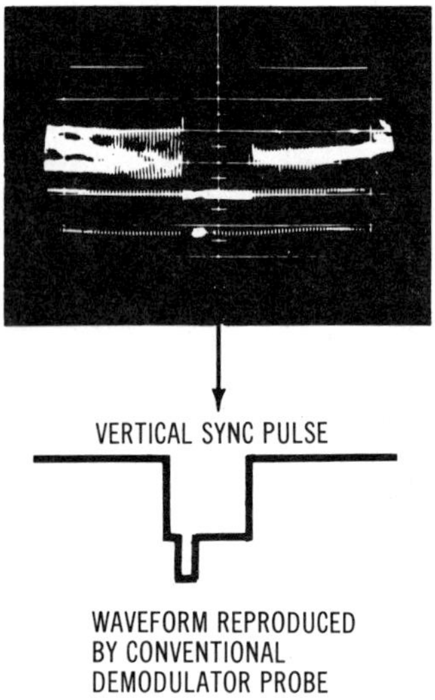

(A) Horizontal sync pulses and equalizing pulses are removed.

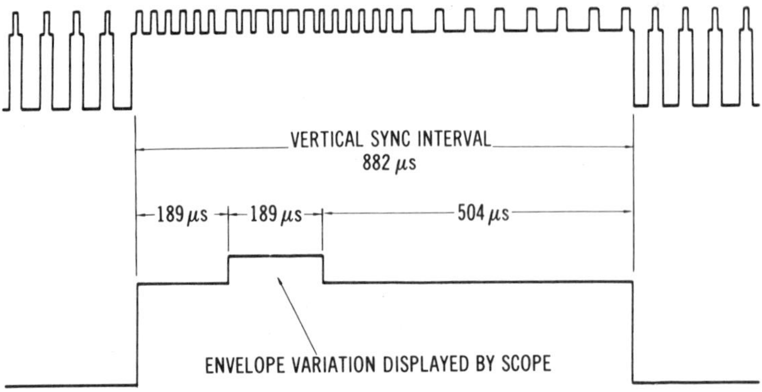

(B) Waveform timing diagram.

Fig. 3-7. Only the lower frequencies are reproduced by a conventional demodulator probe.

the input terminals of the second oscilloscope. Connect one of the demodulator probes to the input of the second if stage, and connect the other demodulator probe at the input to the picture detector.

Procedure: Tune in a station signal (or use a pattern generator), and adjust the receiver controls for normal operation. Adjust the oscilloscope(s) to display the signals at the monitored points. If the intermittent does not occur within a reasonable time, it can often be speeded up by tapping circuit boards,

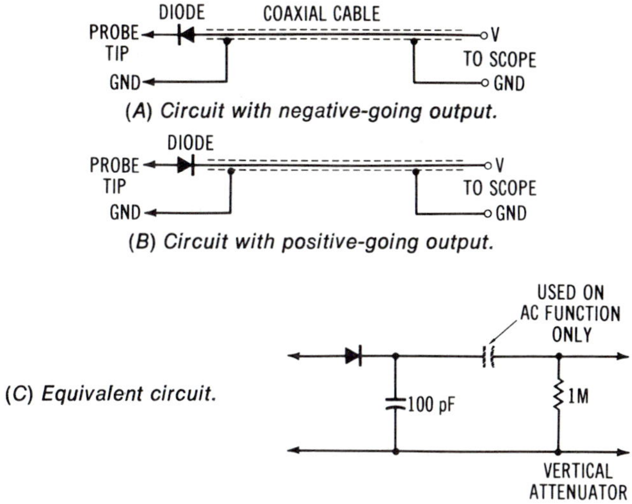

(A) Circuit with negative-going output.

(B) Circuit with positive-going output.

(C) Equivalent circuit.

Fig. 3-8. Simplest demodulator probe design (series detection).

devices, and components, by flexing leads, by heating and cooling suspected parts, or by varying the line voltage. Sometimes the intermittent will occur if the receiver is switched on and off several times in succession.

Evaluation of Results: When the signal is suddenly blocked, or attenuated, this fact becomes evident on one or both channels of the oscilloscope. It is helpful to use two dual-trace scopes, because more signal points can be monitored simultaneously, thereby helping to pinpoint the intermittent. Monitoring is the most practical approach to tracing intermittents, because transfer of a probe from point to point, as in ordinary signal tracing procedures, causes transients which frequently trigger the intermittent, thereby defeating the analytical process.

NOTE 3–3

Medium Impedance Demodulator Probe

A medium-impedance demodulator probe, such as depicted in Fig. 3-10, is preferred by some technicians because it provides a reasonably good display of the horizontal sync pulse, as well as the vertical sync pulse. The chief disadvantage of the medium-impedance probe is the increased circuit loading that it imposes. When several probes are simultaneously applied in the signal channel, loading may be a serious "side effect" unless an appropriate technique is used. Capacitive coupling is helpful in reduction of loading; instead of applying a demodulator probe directly at a test point, one or more layers of masking tape can be inserted. Although the signal level is attenuated, sufficient pickup is obtained for practical testing at the higher-level points.

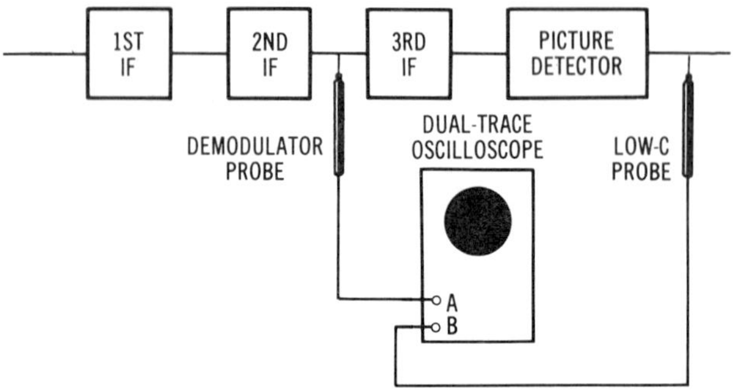

(*A*) *A dual-trace scope monitoring the output of the second if stage and output of picture detector.*

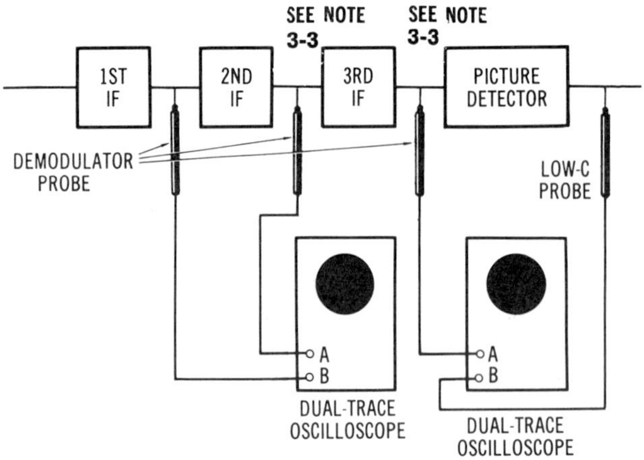

(*B*) *Two dual-trace scopes monitoring four signal points in the if section.*

Fig. 3-9. Dual-trace scopes can monitor two or more signal points simultaneously.

3–6. To Measure the Gain of a TV IF Stage

Equipment: An am signal generator, agc override dc source.

Connections Required: Connect scope to the picture-detector output via a lo-C probe, as shown in Fig. 3-11. Connect a bias box between the agc line and ground. Apply output from the generator at the output, and then at the input of the stage under test.

Procedure: If the generator does not have a series output capacitor, insert a capacitor as shown to avoid drain-off of bias voltage. Set generator to center frequency of if channel, and use amplitude-modulated output; 30 percent modulation is

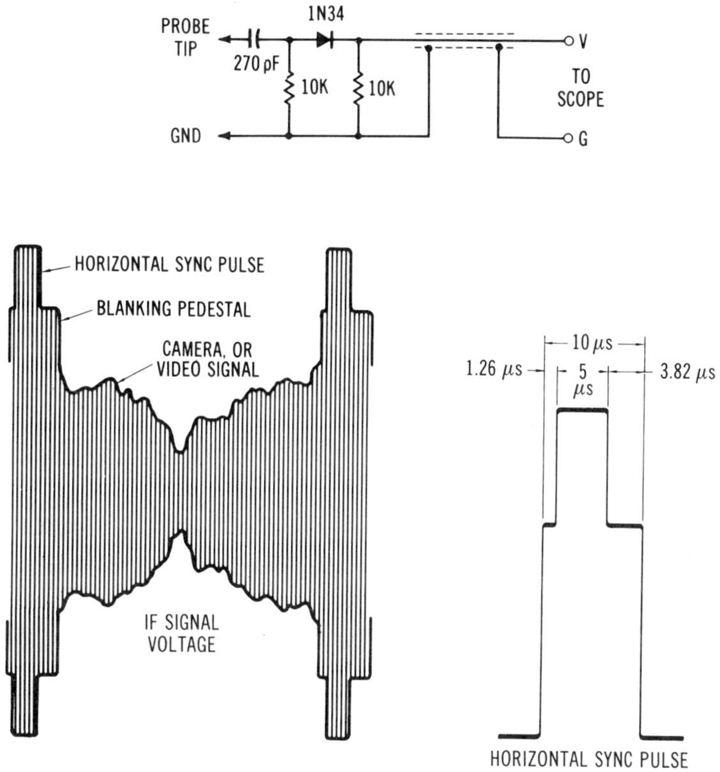

Fig. 3-10. A medium-impedance demodulator probe that can reproduce horizontal sync pulses.

adequate. Adjust override bias to normal agc value, as specified in the receiver service data. As the generator signal is transferred from the output to the input of a stage, observe the change in pattern height on the scope screen. Do not overload a stage—back off on the generator output level, as required. Overload is evidenced as compression or clipping (flat-topping) of the displayed sine wave.

Evaluation of Results: The voltage gain of the stage is given by the ratio of pattern heights when the generator injection point is changed from the output to the input of the stage. The gain of an agc-controlled stage is not fixed, but will change in accordance with the value of override bias that is used. Note that unless the bias is clamped by a dc source, the agc level will change as the signal-injection point is transferred, and the result will be in appreciable error.

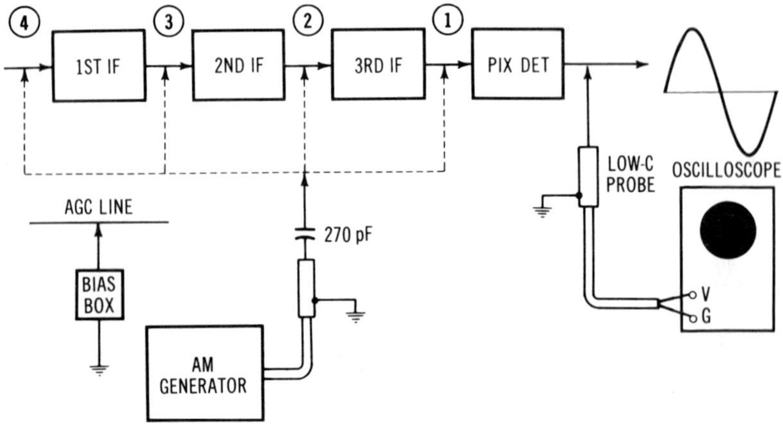

Fig. 3-11. Gain of third if stage is measured by transferring signal-injection point from (1) to (2).

NOTE 3–4

Series and Shunt Detector Configurations

Both series and shunt detector configurations are used in probes and in signal detectors (Fig. 3-12). An occasional demodulator arrangement

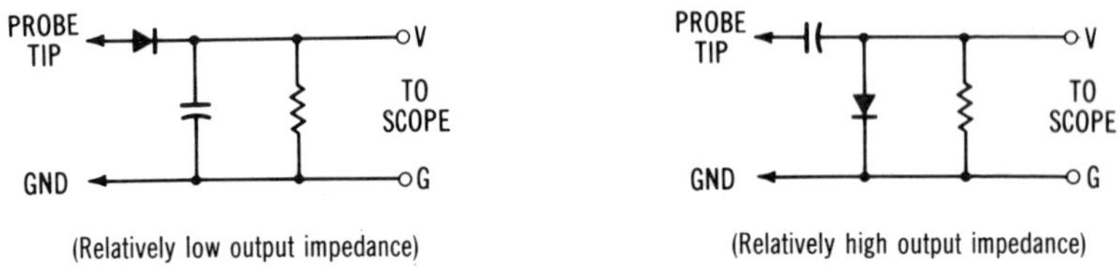

(*A*) *Series detector.* (*B*) *Shunt detector.*

Fig. 3-12. Basic detector arrangements.

uses combination series/shunt detection. The chief distinction between series and shunt detection is that the latter has comparatively high output impedance. Maximum signal power is transferred when the source impedance matches the load impedance.

3–7. To Check the Overall RF/IF Response Curve in a TV Receiver

Equipment: Sweep-and-marker generator, resistive "isolating" probe, agc clamp voltage.

Connections Required: Connect output from generator to antenna-input terminals of receiver; connect bias box between agc line and ground; connect scope to picture-detector load resistor via resistive "isolating" probe, as shown in Fig. 3-13.

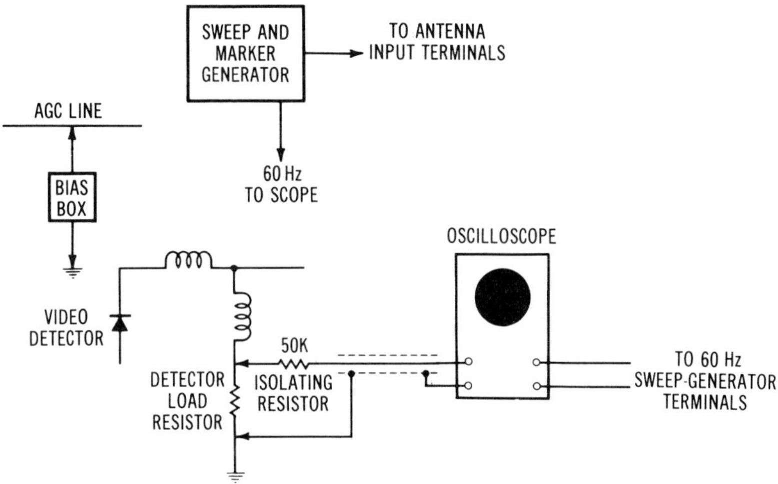

(*A*) *Test connections for check of overall rf/if response curve.*

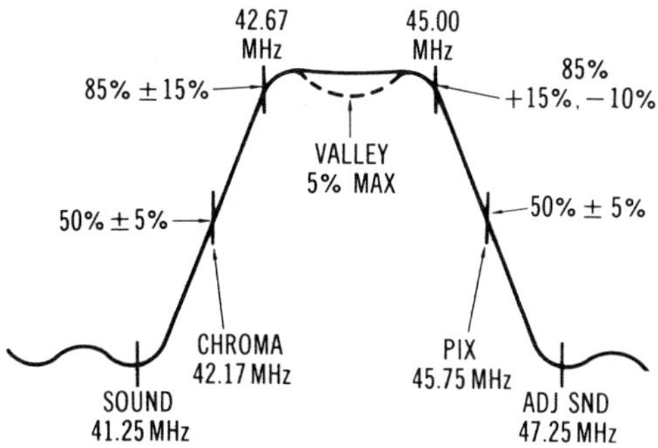

(*B*) *Standard response, showing valley tolerance.*

Fig. 3-13. Application of resistive "isolating" probe in sweep alignment.

Procedure: Set generator for sweep-if and marker output; observe any procedural notes provided in the receiver service data; operate scope on 60-Hz sine-wave horizontal-deflection voltage from sweep generator; adjust output level from generator to avoid overload and resulting artificial flat-topping of displayed response curve.

Evaluation of Results: Check displayed response curve against specified curve in receiver service data. A standard overall response curve is depicted in the accompanying diagram;

the bandwidth is 3.58 MHz. Black-and-white receivers may utilize somewhat less bandwidth.

NOTE 3–5

Injection of IF Sweep-and-Marker Signal

In normal operation of a receiver, the rf/if frequency response curve will have practically the same shape as the if response curve, since most of the gain and selectivity is developed by the if amplifier. To check the if response, the same test setup is used as above, except that the if sweep-and-marker signal is injected at the base of the mixer transistor, as shown in Fig. 3-14. Check the receiver service data for any special instructions; for example, directions may be given for disabling the local oscillator, for use of an emitter-injection point, or for a two-part procedure to check the if response.

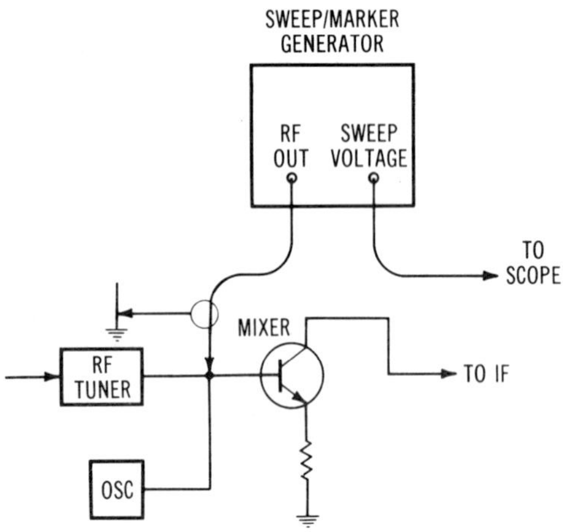

Fig. 3-14. Standard signal-injection point for checking the if frequency response.

3–8. To Check the RF Response Curve in a TV Receiver

Equipment: Same as in Fig. 3-1.

Connections Required: Same as in Fig. 3-1, except that the scope takeoff point is made at the mixer transistor (a "looker" point in the mixer base return circuit may be specified in the receiver service data).

Procedure: Set generator for vhf sweep-and-marker output; observe any procedural notes that may be provided in the receiver service data; operate scope on 60-Hz sine-wave horizontal-deflection voltage from sweep generator.

Evaluation of Results: As exemplified in Fig. 3-15, the frequency

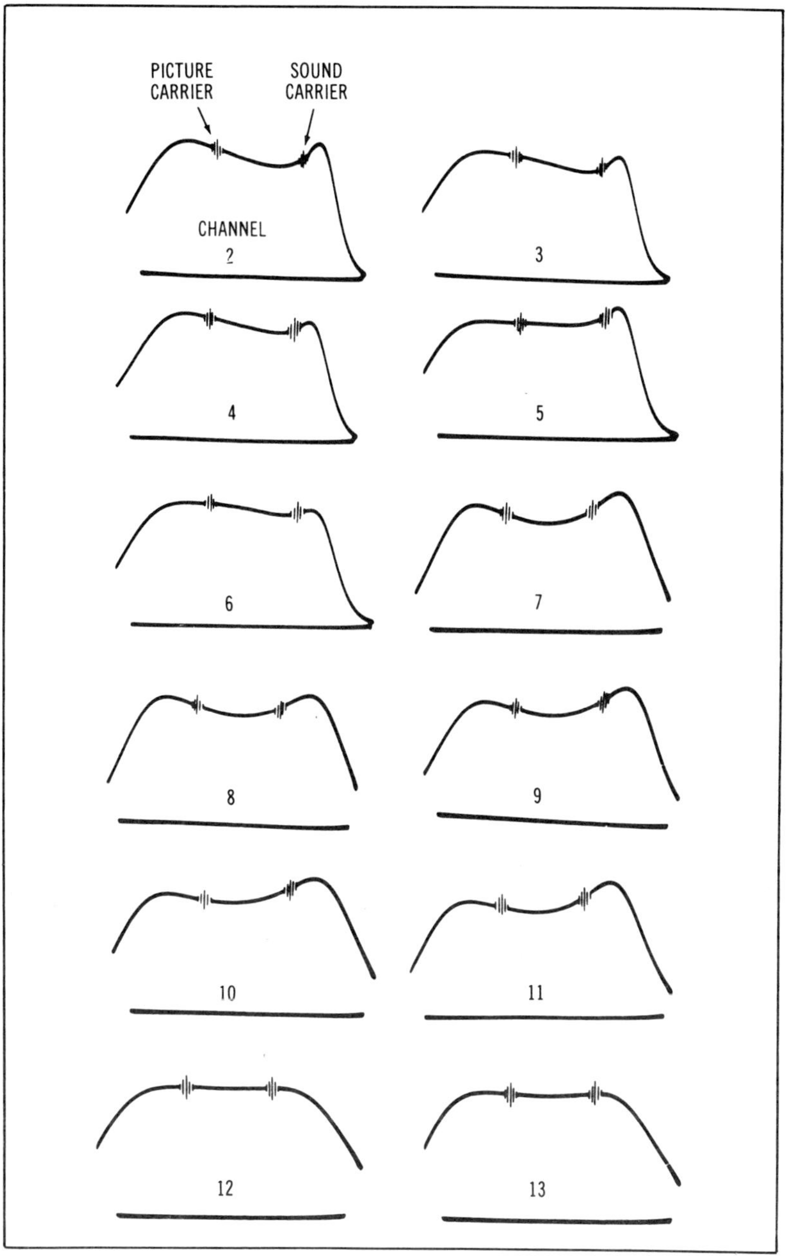

Fig. 3-15. The rf response curves showing normal tolerances on the 12 vhf channels for a good-quality receiver.

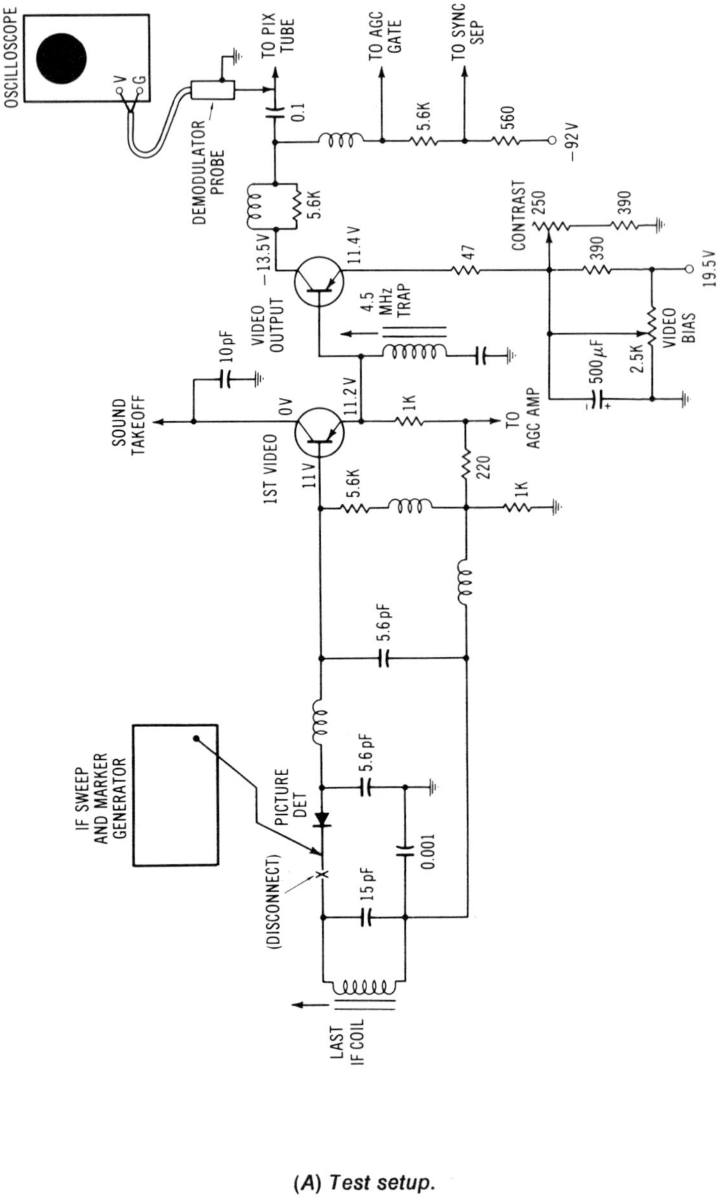

(*A*) *Test setup.*

Fig. 3-16. Check of video amplifier

response of each channel should be checked. Although successive channels will not have identical response curves, a reasonable tolerance includes both the picture carrier and the sound carrier on top of the curve, absence of substantial sag, and absence of sharp and high peaks. When the response curve cannot be contoured satisfactorily by adjustment of the alignment trimmers and slugs, it is indicated that troubleshooting is required. Most shops prefer to send defective tuners to specialized repair depots.

3–9. To Check the Video-Amplifier Frequency Response

Equipment: Sweep-and-marker generator, demodulator probe.

Connections Required: Disconnect picture-detector diode from the last if coil, and apply sweep-and-marker signal through the picture-detector diode to the input of the video amplifier, as shown in Fig. 3-16. Connect scope via demodulator probe to output of video amplifier.

Procedure: Use if output from the sweep generator, and adjust the marker frequency to obtain a video-frequency response curve with zero frequency at the left-hand end of the base line, as exemplified below. The 4.5-MHz trap dip will normally be displayed toward the right-hand end of the base line. Usually, maximum output is required from the generator, and the scope is operated at fairly high gain.

Evaluation of Results: Although this basic method of testing does not provide markers on the video-amplifier frequency-response curve, the 4.5-MHz trap dip serves as a guide and

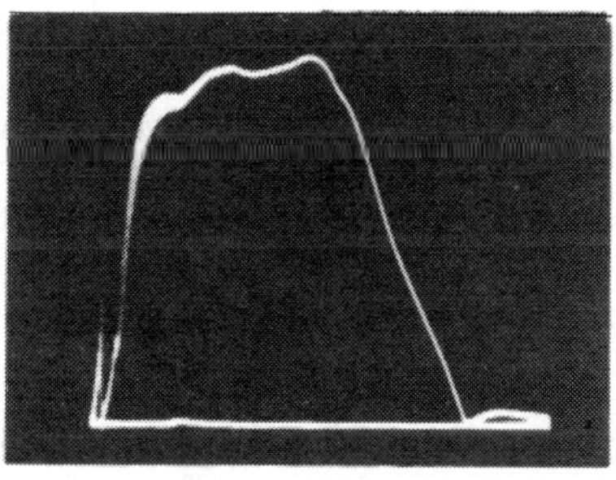

(B) Typical video amplifier response curve.

frequency response.

permits the technician to determine whether the bandwidth is seriously subnormal, and if there are abnormal peaks or suck-outs in the response curve. If an absorption-marker box is connected in series with the "hot" lead of the demodulator probe, "dip" markers will be displayed at specified frequencies along the response curve.

Alternate Method: If an elaborate sweep-and-marker generator is used which provides a video-frequency sweep output, the test signal is then applied directly at the input of the video amplifier (not through the picture-detector diode). This type of sweep generator usually has built-in absorption-marker facilities.

NOTE 3–6

Formation of Video-Frequency Test Signal

When an if sweep-and-marker signal is used to check the video-amplifier frequency response, the if test signal is passed through the picture-detector diode in order to heterodyne the two frequencies and thereby develop a zero- to 5-MHz sweep signal. In other words, the video-frequency test signal is the difference frequency between the if sweep and marker signals. Note that absorption markers are generally preferred for marking a video-frequency response curve in order to avoid interharmonic beats with resulting spurious markers.

3–10. To Check the Overall Picture-Channel Response With a VITS Signal

Equipment: None.

Connections Required: Connect oscilloscope at output of video amplifier (or at output of picture detector) via lo-C probe.

Procedure: Tune in a color-tv station signal, and adjust receiver controls for normal reception. Adjust triggered-sweep scope controls to display the multiburst VITS signal. Then, adjust scope to display the next horizontal line with the sine-squared pulse.

Evaluation of Results: As shown in Fig. 3-17, the VITS multiburst signal is transmitted with uniform amplitude from 0.5 MHz to 4.2 MHz. After passage through the receiver circuits, the bursts will become attenuated in accordance with the overall frequency response of the picture channel. In turn, a helpful

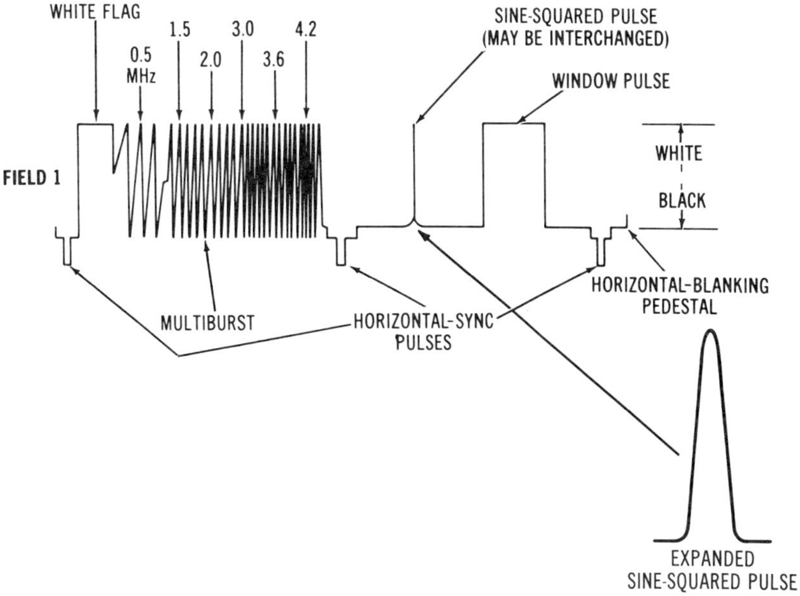

(*A*) *First VITS scanning line.*

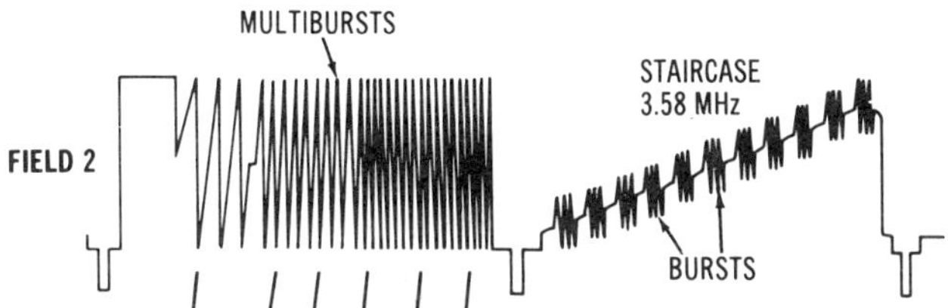

(*B*) *Second VITS scanning line.*

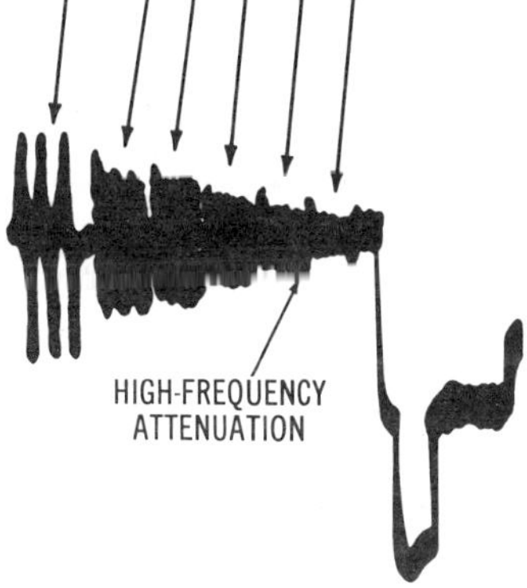

NOTE: These waveforms are transmitted by color-TV stations on two lines of the vertical-retrace interval.

(*C*) *Distortion of multiburst signal.*

Courtesy Sencore, Inc.

Fig. 3-17. Vertical-interval test signal (VITS).

quick-check is provided; if the probe is transferred to the picture-detector output, a comparative evaluation of video-amplifier frequency response is provided. Next, as shown in Fig. 3-18, the sine-squared (T pulse) waveform is transmitted without undershoot. After passage through the receiver circuits, the pulse will normally undershoot with one leading lobe and one trailing lobe. Additional lobes indicate abnormal frequency response. Unsymmetrical leading and trailing undershoots indicate nonlinear phase response.

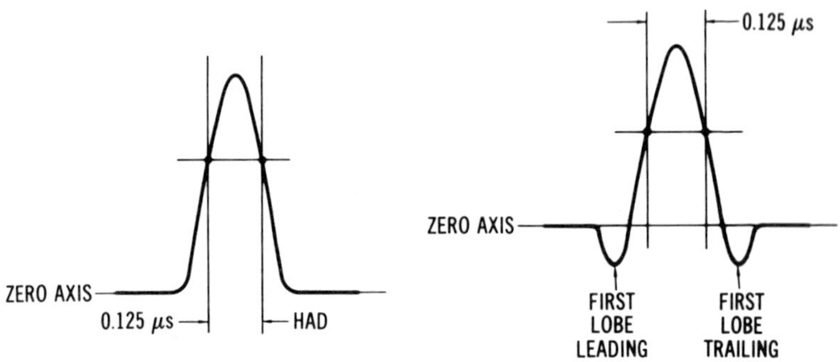

(A) Transmitted waveform with half-amplitude duration (H.A.D.) of 0.125 μs.

(B) Normal reproduction of T pulse by receiver circuits.

Fig. 3-18. T pulse waveforms.

NOTE 3–7

Sine-Squared Pulse and Picture Element

The sine-squared pulse with a half-amplitude duration of 0.125 μs (T pulse) is very informative in tv test procedures because the pulse is practically the same as one picture element. In other words, picture elements will be distorted by the receiver circuits to the same extent that a T pulse is distorted. Note also that the scope used in this test should have better T-pulse response than a normally operating tv receiver.

3–11. To Make a Square-Wave Check of the Video Amplifier

Equipment: Square-wave generator.

Connections Required: As shown in Fig. 3-19, connect the output from a square-wave generator to the input of the video amplifier (disconnect video-detector diode from input of video amplifier). Connect scope via lo-C probe at the output of the video amplifier.

Procedure: Adjust square-wave generator for an output of approximately 1 V p-p, at a repetition rate of 100 kHz. Set contrast control of receiver to its normal operating position.

Evaluation of Results: A reasonably undistorted 100-kHz square wave will be reproduced by a normally operating good-quality receiver. As exemplified in Fig. 3-19B, video amplifiers are

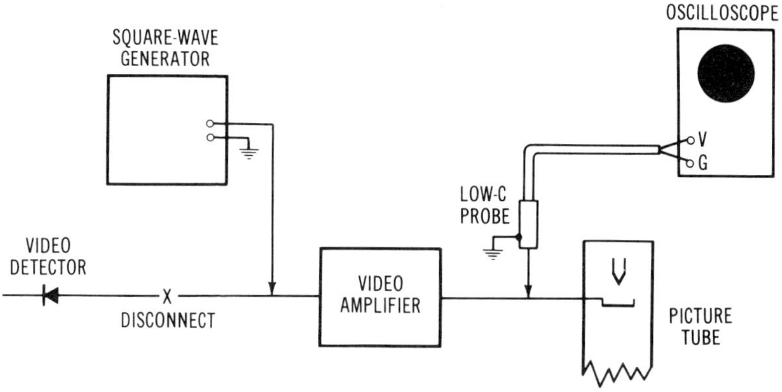

(A) Test setup.

(B) Normal 100-kHz square-wave response.

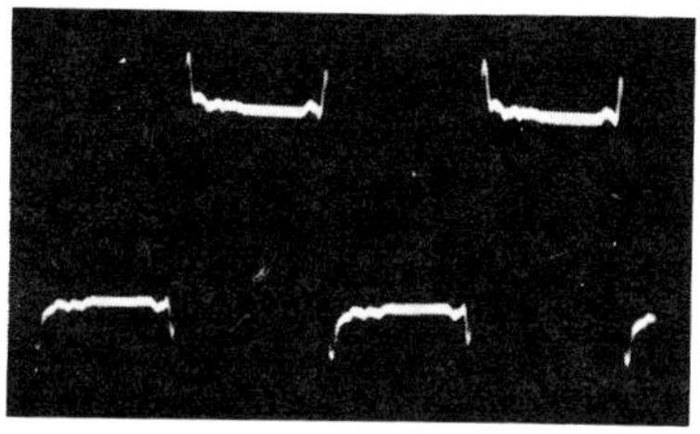

Fig. 3-19. Checking square-wave response of a high-performance video amplifier.

often designed to provide a small amount of overshoot. This is done to minimize the rise time of the waveform, and also to provide accented edge transition ("crispening") of the image. Subnormal rise time and excessively rounded corners in the reproduced square wave indicate inadequate frequency response, often due to capacitor defects.

NOTE 3–8

Rise Time Versus Cutoff Frequency

Rise time is measured from the 10 percent point on the leading edge to the 90 percent point, as shown in Fig. 3-20. A triggered-sweep oscilloscope must be used, and the vertical amplifier of the scope must have

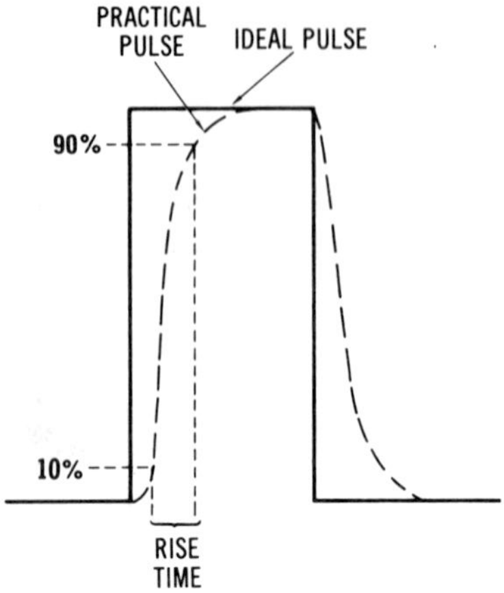

Fig. 3-20. Measuring rise time.

greater bandwidth than the video amplifier under test. Note that the rise time of the reproduced square wave is related to the frequency response of the video amplifier as follows:

$$f_{co} = \frac{1}{3T_r}$$

where,

f_{co} is the amplifier's cutoff frequency,
T_r is the amplifier's rise time.

For example, if a video amplifier has a 4-MHz cutoff frequency, its square-wave rise time will be 0.08 microsecond.

3–12. To Signal-Trace a Sync-Buzz Pulse

Equipment: None, although a pattern generator may be helpful.

Connections Required: None.

Procedure: With a lo-C probe, start signal tracing at the primary of the ratio-detector transformer, and proceed back through the intercarrier-if amplifier and the video amplifier. As shown in Fig. 3-21, a buzz pulse often appears as downward modulation of the 4.5-MHz intercarrier signal by a stripped vertical-sync pulse. Operate the scope on 30-Hz sweep.

Evaluation of Results: The source of a sync-buzz pulse is often in an overloaded video-amplifier stage. In such a case, the scope will display a pulse at the malfunctioning stage, but not at any preceding stage. In marginal situations, the buzz pulse will be most prominent when a white-background signal is present without any camera signal. Overload commonly re-

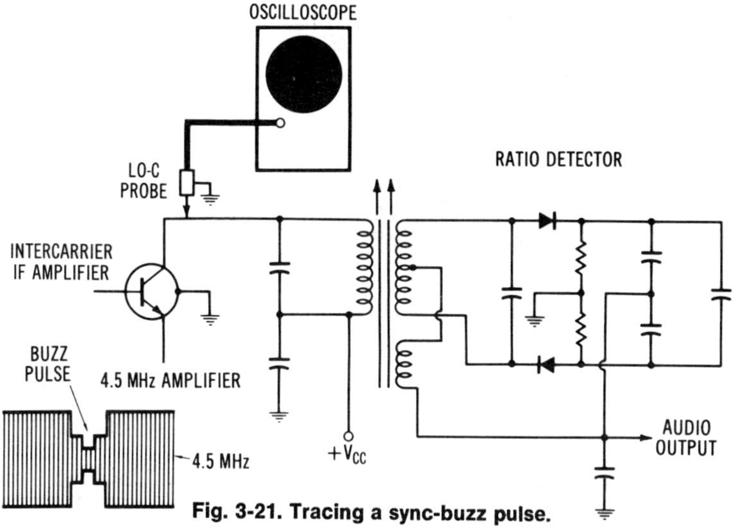

Fig. 3-21. Tracing a sync-buzz pulse.

sults from incorrect bias voltage, or collector leakage in a transistor.

NOTE 3–9

Sync Buzz Versus Sweep Buzz

Although less common, sync buzz can also arise in an if amplifier stage. Misalignment, in which the sound carrier rides too high on the frequency response curve, is the most probable cause. The intensity of the synz buzz is increased when nonlinear amplification (overload) occurs. *Sync buzz should not be confused with sweep buzz. If the vertical-sweep pulse gains entry into the sound channel, 60-Hz sweep buzz will occur. Sweep buzz sounds similar to sync buzz. However, when the vertical-hold control is turned to roll the picture, sweep buzz will change in tone, whereas sync buzz will not change its tone.*

Note also that an intercarrier-if amplifier can be signal traced by using a service-type am generator as a signal source. Although the fm detector and limiter reject amplitude modulation, a service-type am generator produces incidental frequency modulation, as depicted in Fig. 3-22. Incidental fm is generally greater at high percentages of amplitude modulation.

3–13. To Check Sync Separator Action

Equipment: None (unless a pattern generator is preferred as a signal source).

Connections Required: Connect oscilloscope via lo-C probe to output of differentiating circuit, and then to output of integrat-

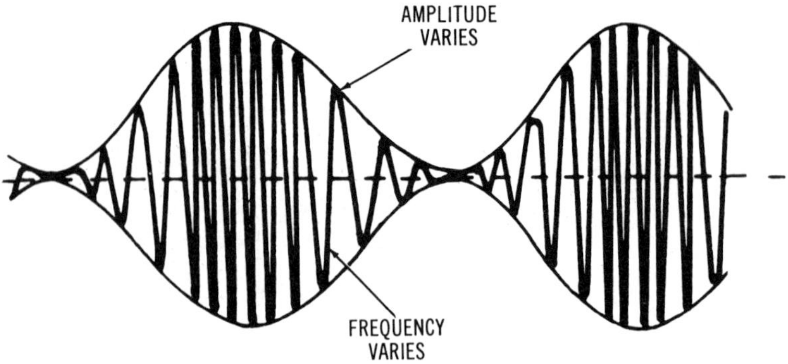

Fig. 3-22. A 100-percent amplitude-modulated output signal from an am generator, with incidental frequency modulation.

ing circuit, as exemplified in Fig. 3-23. Connect output from pattern generator (if used) to antenna-input terminals of receiver.

Procedure: Adjust receiver controls for normal operation.

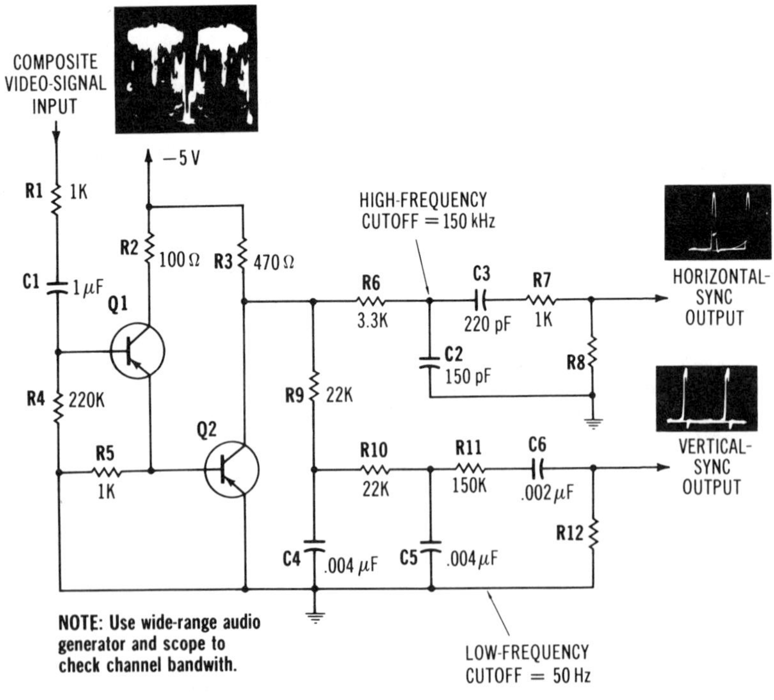

Fig. 3-23. A widely used sync-separator arrangement.

Evaluation of Results: The output waveform from the differentiating circuit should be similar to that specified in the receiver service data, and should have an amplitude within ±20 percent of the specified value. In particular, the differentiator output should be practically free from residual camera signal. The output waveform from the integrator should have a shape similar to that specified in the service data, and an amplitude within ±20 percent of the specified value. (Some receivers have a ±10 percent amplitude specification.)

NOTE 3–10

Bandwidth of Horizontal Sync Channel

A horizontal sync channel should have a bandwidth of 135 kHz, from 15 kHz to 150 kHz. This bandwidth provides maximum reduction of noise, without undue sync-pulse attenuation. Bandwidth is controlled by RC values in the sync channel. A vertical sync channel normally has a lower frequency-response limit of 50 Hz to ensure rejection of possible low-frequency interference. Sync-pulse tolerances of ±10 percent and of ±20 percent are indicated in Fig. 3-24. Observe that the sync-separator circuit has a clipping and limiting action which tends to keep the output pulse amplitude constant as the input signal amplitude varies.

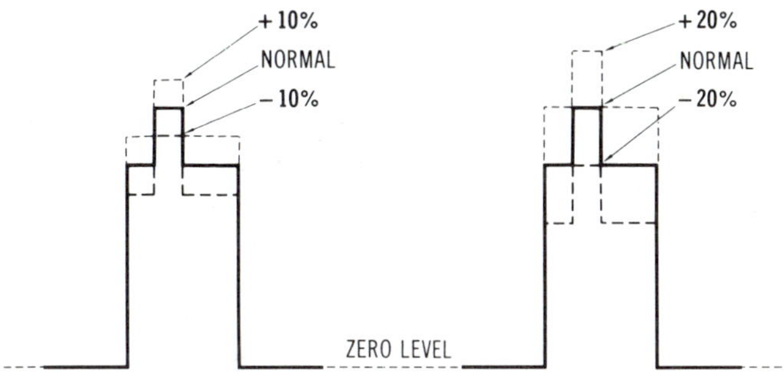

Fig. 3-24. Sync pulse amplitude tolerances.

3–14. To Check an AFC Section (Ringing Circuit Type)

Equipment: None.

Connections Required: Apply scope via lo-C probe at base of horizontal-oscillator transistor.

Procedure: Adjust receiver controls for normal operation.

Evaluation of Results: With reference to Fig. 3-25, observe the leading edge of the base waveform. In normal operation, the rise is rapid into the conduction region of the transistor (near

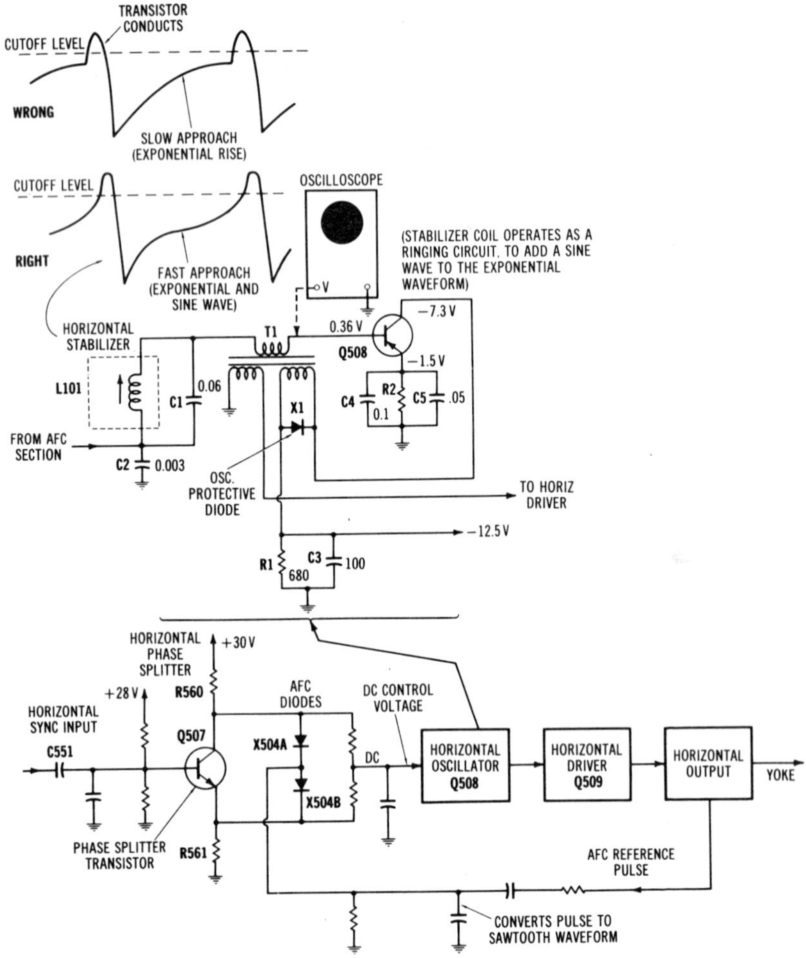

Fig. 3-25. The afc section is a portion of the servo (feedback control system).

the peak of the waveform). A slow approach to the conduction region results in poor noise immunity. Adjust the horizontal stabilizer coil to obtain the maximum rate of change as the leading edge approaches the peak of the base waveform.

NOTE 3–11

Faulty Sync Locking Action

The dc control voltage for a ringing-circuit afc arrangement is derived from a pair of afc diodes, as exemplified in the diagram. These diodes operate in a balanced-bridge configuration, which is part of the afc servo system. Unless the diodes are closely matched, the bridge will operate in an unbalanced state, and horizontal locking action will be impaired. Unstable sync lock also results from leaky capacitors in the afc section. Collector-junction leakage in the phase-splitter transistor reduces the dynamic range of the control system and causes "touchy" sync locking action.

3–15. To Check the Drive Waveform to the Horizontal-Output Stage

Equipment: None.

Connections Required: Apply scope via lo-C probe at base of horizontal-output transistor, as shown in Fig. 3-26.

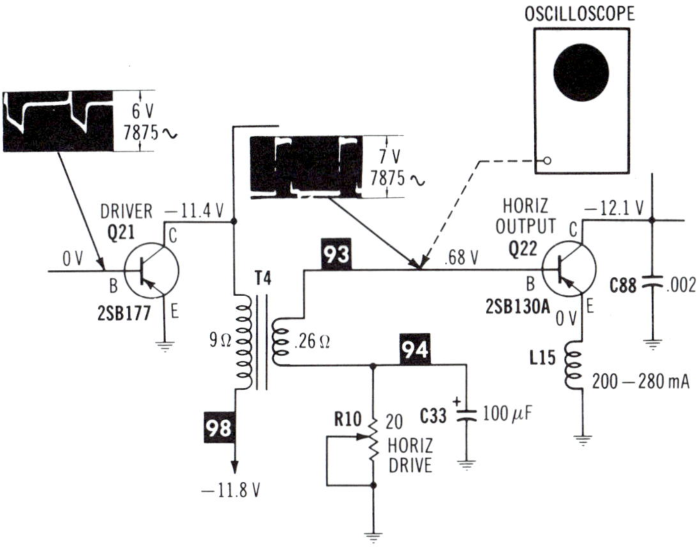

Fig. 3-26. Drive waveform must have normal amplitude and fast rise.

Procedure: Observe amplitude of drive pulse. Adjust horizontal-drive control, if necessary, to obtain specified peak-to-peak voltage. Measure the rise time of the drive pulse.

Evaluation of Results: In case that adequate driving amplitude cannot be obtained, check driver input waveform (Fig. 3-26). A subnormal driver input waveform may result from a defect in the horizontal-oscillator circuit, or from collector junction leakage in the driver transistor. Note that although the drive waveform to the horizontal-output transistor has normal amplitude, the output transistor will overheat unless the drive waveform has a sufficiently fast rise time—make a comparison test with a receiver in good working condition, in case of doubt. Slow rise time generally points to capacitor defects.

3–16. To Check the Operation of a Vertical Sweep System

Equipment: None.

Connections Required: Apply scope via lo-C probe to the input and output terminals of the oscillator, driver, and output terminals progressively.

Procedure: Observe the pattern waveshapes and amplitudes, with receiver controls adjusted for normal operation.

Evaluation of Results: With reference to Fig. 3-27, the waveshapes should agree reasonably well with those specified in the receiver service data, and the waveform amplitudes should fall within ±20 percent of the specified values. Note that nonlinearity in this type of vertical-sweep system is often the result of failing electrolytic capacitors. Because of stage interaction, it is impractical to localize a marginal electrolytic capacitor on the basis of scope tests alone.

3–17. To Check the Color Burst in the Complete Color Signal

Equipment: None (unless a color-bar generator is preferred as a signal source).

Connections Required: Connect oscilloscope via lo-C probe to output of picture detector or at a point in the video amplifier prior to the color-subcarrier trap. Connect output from color-bar generator (if used) to antenna-input terminals of receiver.

Procedure: A dual-trace scope is preferable; triggered sweep is necessary. With the complete color signal displayed on one channel, adjust time-base controls to display the color-burst waveform on the other channel; advance vertical gain control for second channel to display the burst at convenient amplitude.

Evaluation of Results: As exemplified in Fig. 3-28, the waveshape of the color burst becomes clearly visible at increased gain on the second channel. Although the burst is transmitted at the same amplitude as the sync tip, the if response is normally 6 dB down at 3.58 MHz, with the result

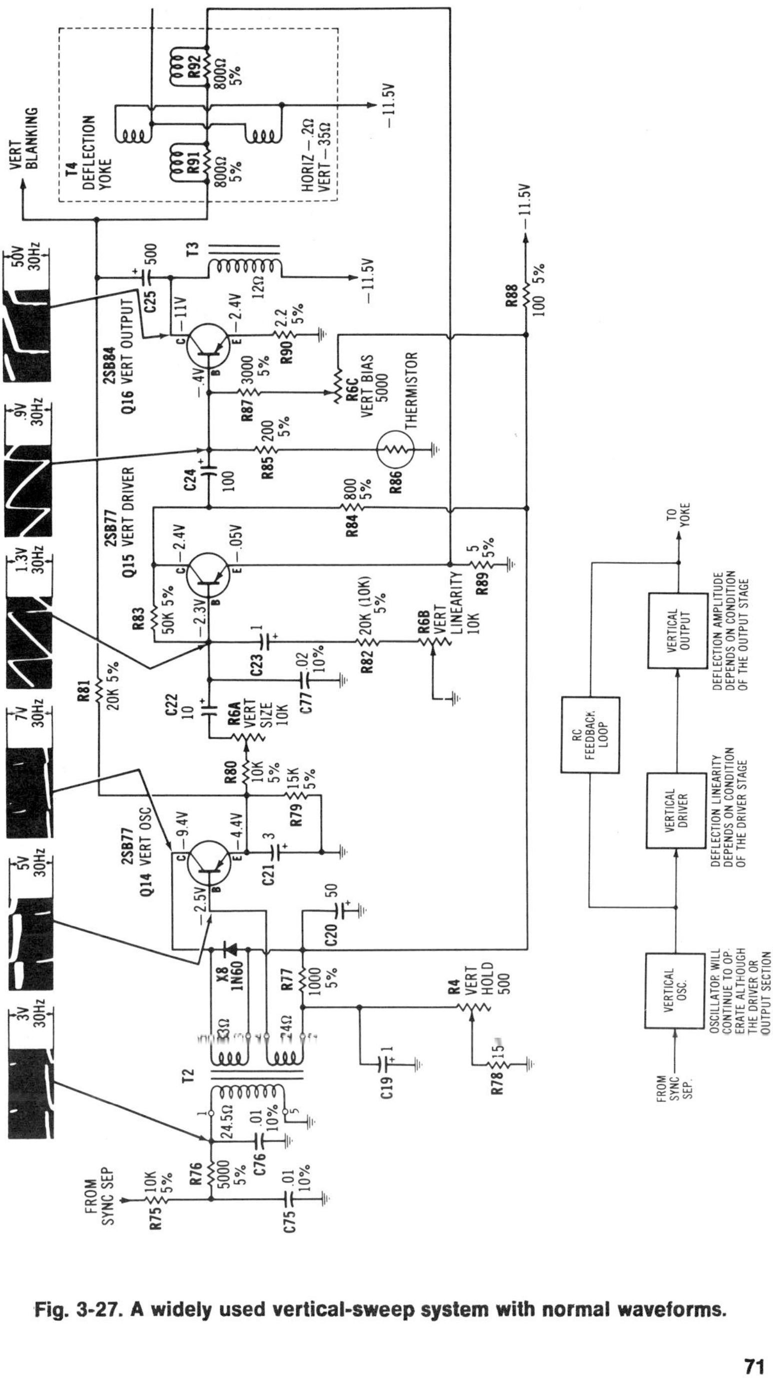

Fig. 3-27. A widely used vertical-sweep system with normal waveforms.

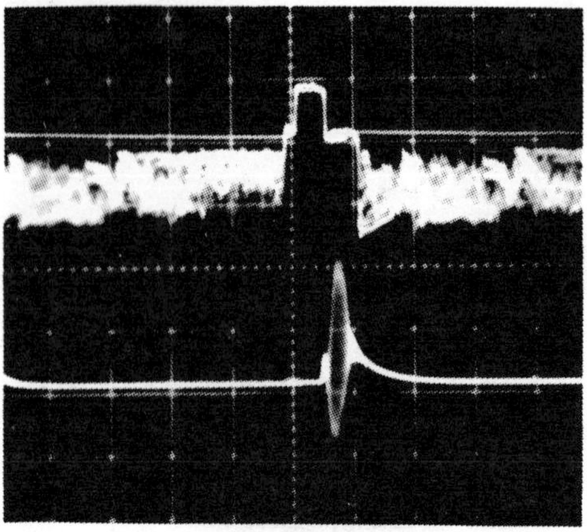

Fig. 3-28. Display of the color burst from the complete color signal.

that the burst undergoes attenuation to one-half the amplitude of the sync tip. In the case of an antenna signal, propagation anomalies and lack of flat frequency characteristic in the antenna system may result in additional attenuation of the burst. If the color burst becomes excessively attenuated, the color-sync section of the receiver cannot operate normally.

NOTE 3–12

Subcarrier-Sound Beat Waveform

When a 920 kHz "ripple" occurs in the video signal pattern, as exemplified in Fig. 3-29, it is indicated that the sound signal and the color subcarrier are heterodyning abnormally, causing a beat frequency to appear in the video waveform. A substantial 920-kHz beat will also be visible as an interference pattern on the picture-tube screen. A scope is helpful in this situation to signal trace the beat interference back to its source. In general, the 920-kHz beat will not be visible if the receiver is properly aligned, and if the color-subcarrier trap is operating normally.

NOTE 3–13

Time Bases in Dual-Trace Oscilloscopes

A dual-trace oscilloscope provides two vertical attenuators; one attenuator operates in the A channel and the other attenuator operates in the B channel. Service-type oscilloscopes generally provide one set of time-base controls; in other words, channels A and B are deflected by the same time base. A more elaborate design provides separate time bases, so that the channel A signal can be deflected at a different rate from the channel B signal. Provisions for delayed triggering of the channel-B time base may be included. With reference to Fig. 3-28 simultaneous display of the complete color signal and of the color burst from the complete color signal requires separate time bases for the two vertical channels.

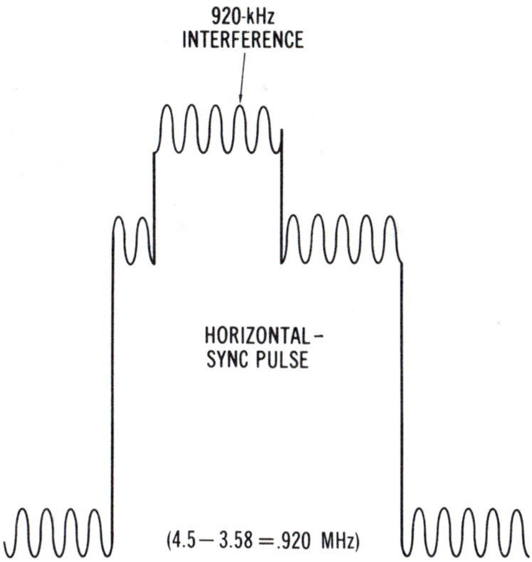

Fig. 3-29. Appearance of 920-kHz interference in the video signal.

3–18. To Check the Operation of the Color Burst Amplifier

Equipment: None (unless a color-bar generator is preferred as a signal source).

Connections Required: Connect the oscilloscope via lo-C probes at the input and at the output of the burst amplifier. Connect output from color-bar generator (if used) to antenna-input terminals of the receiver.

Procedure: Adjust vertical-gain controls for convenient pattern height on each channel. Both channels are deflected from the same time base in this type of display.

Evaluation of Results: As shown in Fig. 3-30, the color burst normally falls on the top of the burst keying pulse. If the burst falls on the side of the keying pulse, it will become attenuated and distorted accordingly. Note that the position of the burst on the keying pulse will move when the adjustment of the horizontal-hold control is varied. In case that the burst is incorrectly positioned when the horizontal-hold control is correctly adjusted, it is most likely that a defect is present in the burst keying circuit. Subnormal amplitude of the burst keying pulse also points to a defective burst-keying circuit.

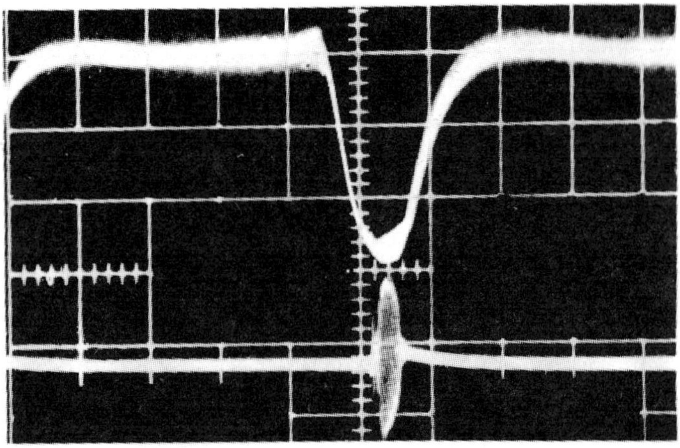

Fig. 3-30. Display of the burst keying pulse with the color burst, and of the separated burst.

NOTE 3–14

Options in Dual-Trace Oscilloscope Functions

As noted previously, a dual-channel oscilloscope may provide A+B and A−B displays, to supplement conventional A and B displays. The more elaborate oscilloscopes may include a vertical delay line, trace alignment (trace-rotation) control, Z-axis modulation facilities, separate trigger controls for channel-A and channel-B time bases, and delayed time-base triggering. A choice of alternate or chopped display modes is often provided. Intensification of the channel-A display is sometimes provided by the delayed time base in order to identify the "window" in the channel-A signal which is being deflected by the channel-B time base. The chart in Fig. 3-31 exemplifies the chief options that are available.

3–19. To Check the Operation of the Chroma Demodulators

Equipment: Keyed-rainbow color-bar generator.

Connections Required: Connect output from generator to antenna-input terminals of receiver. Connect oscilloscope via lo-C probe in turn to the outputs of the chroma demodulators, as shown in Fig. 3-32.

Procedure: Adjust receiver for normal operation; adjust scope controls for convenient display of the demodulated waveform.

Evaluation of Results: In an R−Y/B−Y/G−Y system, the R−Y waveform normally nulls on the sixth pulse; the B−Y waveform normally nulls on the third and ninth pulses; the G−Y waveform normally nulls on the first and seventh pulses. Lack

BANDWIDTH (MHz)	SENS (mV/DIV)	TRACES	DISPLAY (cm)	DISPLAY: ADD/SUB	DELAY LINE	UNCAL LAMPS	VERT MAG	HORIZ MAG	Z-AXIS MOD	TRACE ROTATION	CH-1/CH-2	TRIGGERING: TV	± SLOPE	AUTO	SINGLE SHOT	DELAYED	HOLDOFF
30	5	DUAL	6.4×8	×	×	×		×10	×	×	×	×	×	×	×	×	×
30	5	DUAL	8×10	×	×	×		×10	×	×	×	×	×	×			×
20	5	DUAL	8×10	×				×5	×	×	×	×	×	×			
20	10	DUAL	8×10	×				×5	×	×	×	×	×	×			
20	10	SINGLE	8×10	×				×5	×	×		×	×	×			
10	1	DUAL	8×10				×5	×5	×	×	×	×	×	×			
10	1	SINGLE	8×10				×5	×5	×	×		×	×	×			
20	2	DUAL	4.8×6	×				×5	×	×	×	×	×	×			
10	10	DUAL	4.8×6	×				×5	×		×		×	×			
10	20	SINGLE	8×10						×								
4	20	SINGLE	4.8×6						×								

Courtesy Leader Instruments, Corp.

Fig. 3-31. Chief options in dual-trace oscilloscope designs.

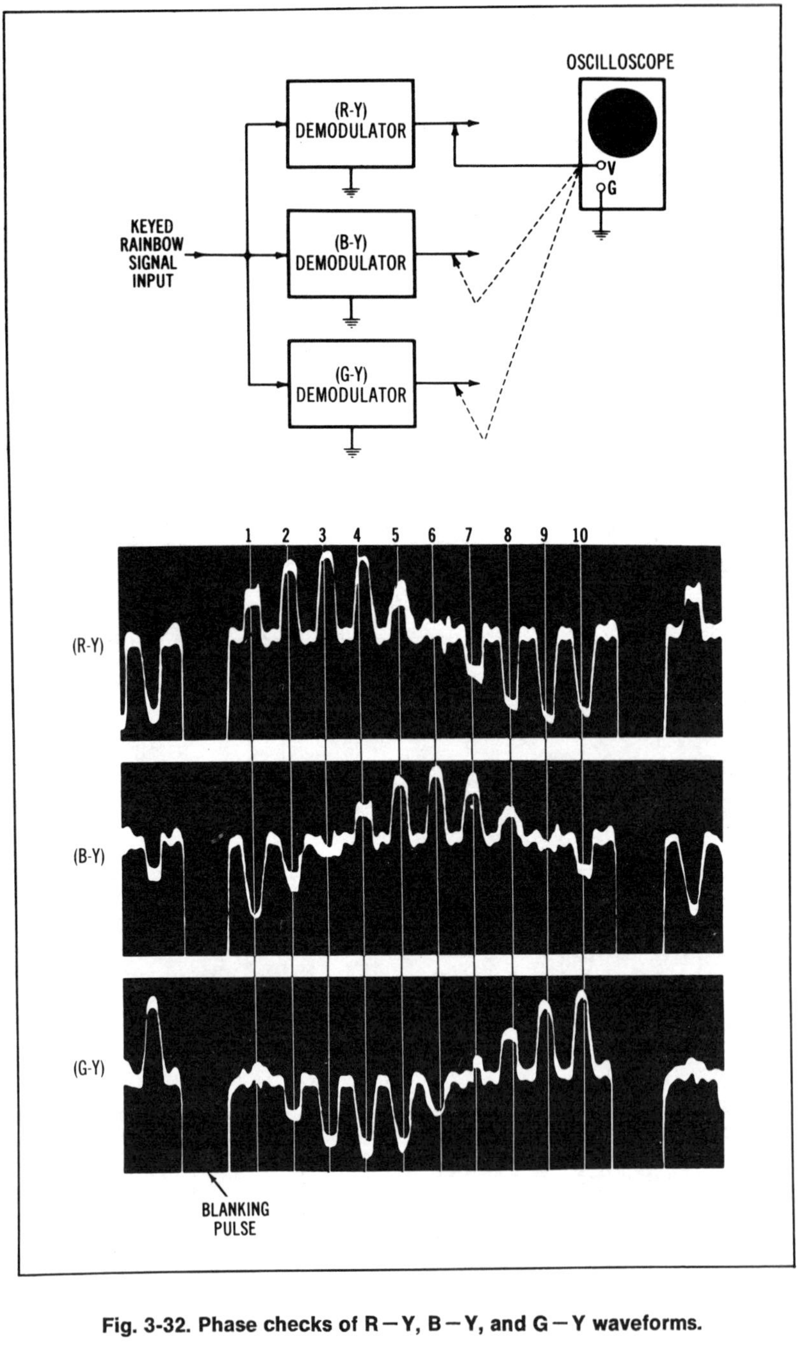

Fig. 3-32. Phase checks of R − Y, B − Y, and G − Y waveforms.

of proper nulls points to a defect in the color-subcarrier phasing circuitry.

NOTE 3–15

Types of Chroma Demodulator Arrangements

Some color receivers use the X/Z demodulator arrangement; others use the RGB configuration. Chroma-demodulation nulls for these designs are somewhat different from those for the R—Y/B—Y/G—Y system. Refer to the receiver service data for the particular receiver. Note also that chroma-demodulation nulls are affected in some receivers by operation of the automatic tint control. In such a case, the waveform nulls should be checked with the tint control turned to the off position. In all receivers, the nulls will shift in accordance with the setting of the manual tint control. Therefore, the tint control should be adjusted to produce correct nulls in the R—Y waveform; in turn, correct nulls will normally be found in the B—Y and G—Y waveforms.

3–20. To Display a Vectorgram

Equipment: Keyed-rainbow color-bar generator.

Connections Required: Connect output from generator to antenna-input terminals of receiver. Connect V and H channels of oscilloscope via lo-C probes to the R—Y and B—Y chroma output terminals of the receiver, as depicted in Fig. 3-33.

Procedure: Adjust receiver controls for normal operation. Adjust vertical and horizontal gain controls of scope for convenient size of pattern.

Evaluation of Results: As exemplified in Fig. 3-33, the displayed vectorgram should approximate a circle. When the tint control is adjusted to make the first "petal" fall at 30 degrees, the third petal normally falls at 90 degrees, and the tenth petal normally falls at 300 degrees. An incorrect phase display points to a defect in the color-subcarrier phasing circuitry.

NOTE 3–16

Distortion in Vectorgram Patterns

A low-level vectorgram is obtained by connecting the oscilloscope at the outputs of the R—Y and B—Y chroma demodulators. A high-level vectorgram is displayed by connecting the oscilloscope at the R—Y and B—Y inputs to the picture tube. If a high-level vectorgram is distorted by one or more "flats," it is indicated that a chroma output amplifier is overloading (Fig. 3-34). This possibility is confirmed by a low-level vectorgram check. However, in case that the demodulator vectorgram is also distorted by one or more "flats," the trouble will be found in the

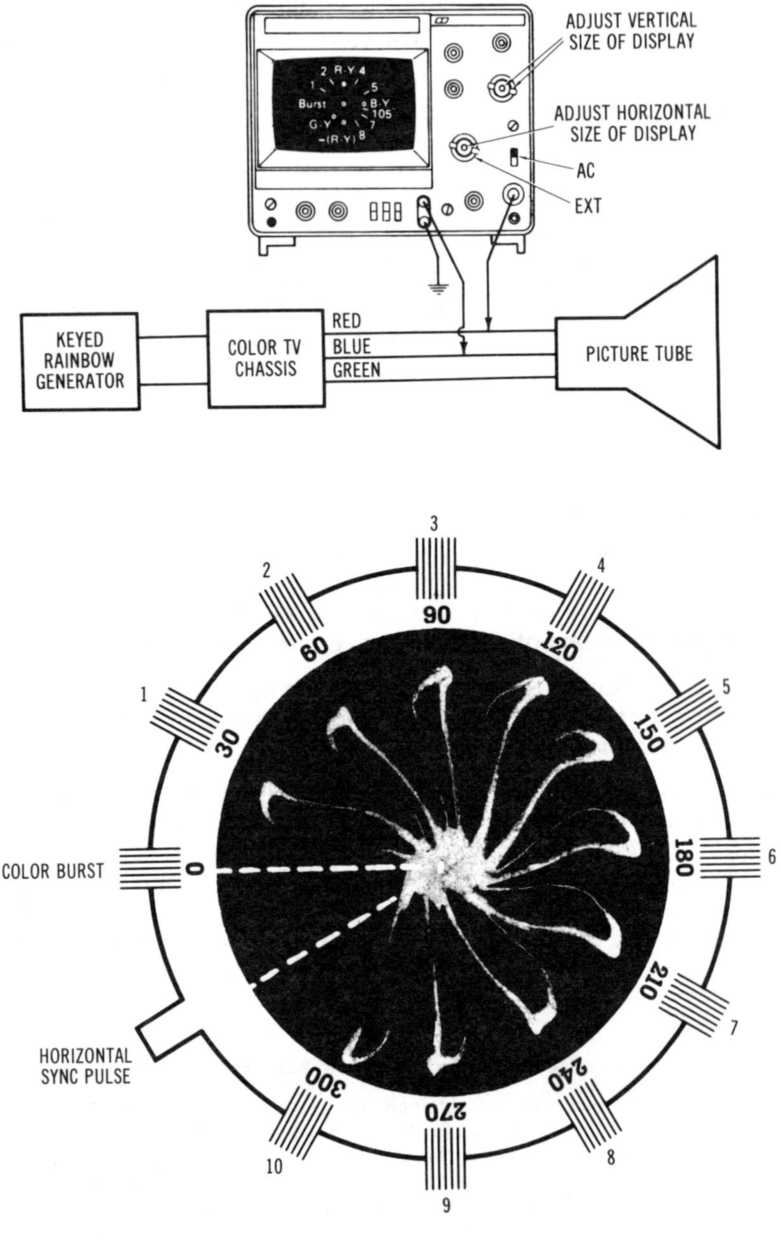

Fig. 3-33. Display of an R – Y/B – Y vectorgram.

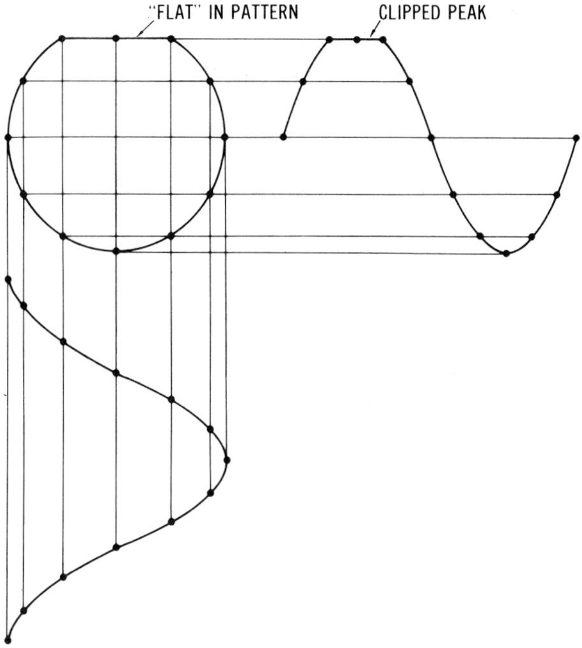

Fig. 3-34. Pattern becomes flattened when one of the input waveforms is clipped.

demodulator circuitry. A demodulator diode is likely to be found defective in this situation.

3–21. To Check the Bandpass-Amplifier Response Curve

Equipment: Video-frequency sweep and marker generator.

Connections Required: Connect output from generator to input of bandpass amplifier per receiver service data. Connect oscilloscope via demodulator probe at output of bandpass amplifier (or as may be specified in the service data).

Procedure: Sweep the bandpass amplifier over a range from 2 to 5 MHz; observe any incidental instructions, such as provision of override bias, that may be given in the service data.

Evaluation of Results: The shape of the frequency response curve should approximate the specified display (the exemplified curve shows a rising high-frequency response), and the markers should fall at the specified points (Fig. 3-35).

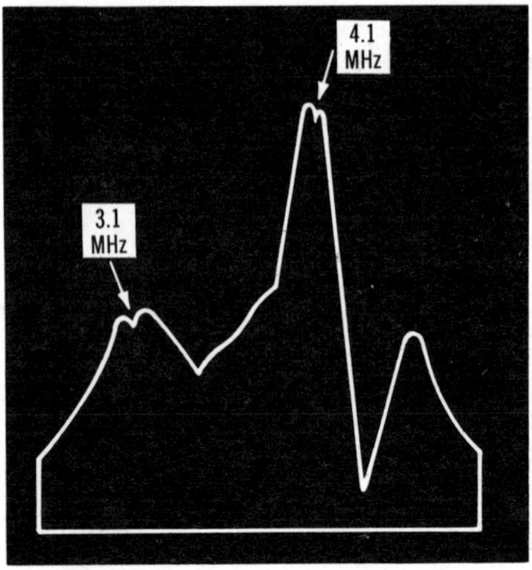

Fig. 3-35. Typical bandpass-amplifier response curve, with absorption markers at 3.1 and 4.1 MHz.

Absorption markers are often used to avoid possible spurious marker production.

3–22. To Make a Video Sweep Modulation Check of a Color Receiver

Equipment: VSM generator, or video sweep generator, marker generator, and absorption marker box.

Connections Required: Connect equipment as shown in Fig. 3-36.

Procedure: Tune marker generator to chosen picture-carrier frequency; adjust video sweep generator for 0- to 5-MHz sweep; adjust scope controls to display pattern as exemplified in Fig. 3-36B. Then, transfer lo-C probe to output of video amplifier; finally, transfer probe to output of bandpass amplifier.

Evaluation of Results: Normally, the outline of the VSM pattern from the picture-detector output is basically the same as the if response of the receiver. The outline of the VSM pattern from the video-amplifier output (in many receivers) is normally flat-topped through 3.58 MHz (check the receiver service data). The outline of the VSM pattern from the bandpass-amplifier output is normally flat-topped from 3.0 to 4.1 MHz in most receivers.

NOTE 3–17

Basic VSM Test Data

VSM patterns may be displayed with use of a lo-C probe, as in Fig. 3-36, or with use of a demodulator probe. When a demodulator probe is used, demodulated response curves are displayed, as shown in Fig. 3-37. In this example, the frequency response at the output of a chroma demodulator is included. Note that a chroma demodulator normally has a bandwidth of 1 MHz. The basic advantage of a VSM test is that it shows how the picture-channel stages operate together as a team. Normally, the end result is to obtain a reasonably uniform frequency response from 3.0 to 4.1 MHz in each of the chroma demodulators.

3–23. To Make a Ringing Test of a Quartz Crystal

Equipment: Pulse generator and lab-type signal generator.

Connections Required: As depicted in Fig. 3-38, connect output from pulse generator to external-modulation terminals of signal generator. Connect quartz crystal in series between the output from the signal generator and the vertical-input channel of the scope.

Procedure: Tune the signal generator precisely to 3.58 MHz; modulate the generator output with a 2 μs pulse. Thereby, the crystal is tested with a simulated color burst. Pulse generator repetition rate should be 15.75 kHz, or less. Observe the resulting display.

Evaluation of Results: As exemplified in the illustration, a normal crystal ringing pattern is almost constant in amplitude; it decays very slowly, because the crystal has a Q value of approximately 8000. If the pattern shows substantial decay over a 60 μs interval, the crystal is defective and should be replaced. Note that the top of the ringing pattern is slightly "wavy," due to the fact that the principal mode of vibration is accompanied by residual minor modes of vibration.

3–24. To Measure the Characteristic Impedance of a Lead-In

Equipment: Vhf sweep generator, assortment of composition resistors.

Connections Required: As shown in Fig. 3-39, connect output from sweep generator to one end of a sample length of the lead-in (such as a 3-foot length). Connect a resistor of suitable value, such as 300 ohms, across the other end of the

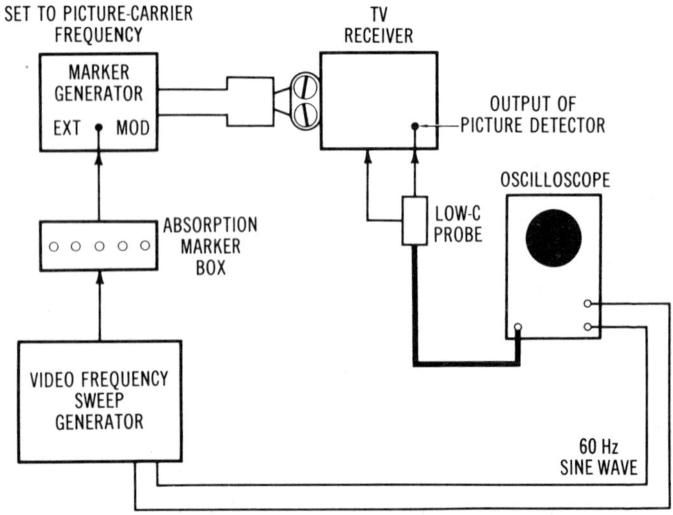

(A) Equipment connections.

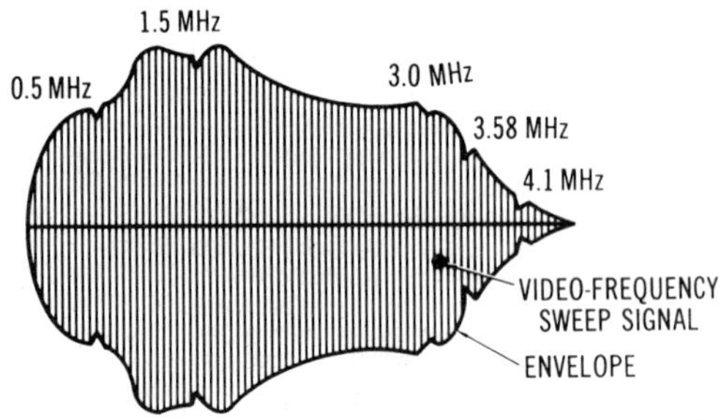

(B) Representative scope pattern at picture detector output.

Fig. 3-36. Typical video-sweep

line. Connect scope via double-ended demodulator probe across the resistor.

Procedure: Set vhf sweep generator to any channel from 2 to 13, with a 10-MHz sweep width. Observe pattern on scope screen, using 60-Hz sine-wave horizontal deflection.

Evaluation of Results: A flat-topped pattern indicates that the line is correctly terminated. The value of the terminating resistor will then be equal to the characteristic impedance of the line. However, if the top of the pattern slopes or is uneven, the value of the terminating resistor is not equal to the characteristic impedance of the line. Substitute other values of resistors until a flat-topped pattern is obtained.

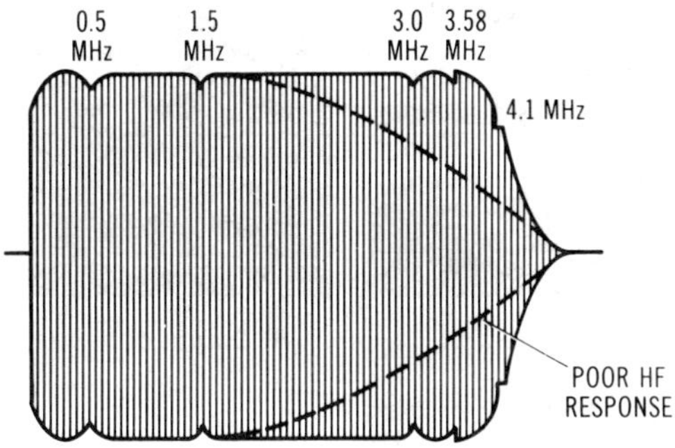

(C) Typical pattern at video amplifier output.

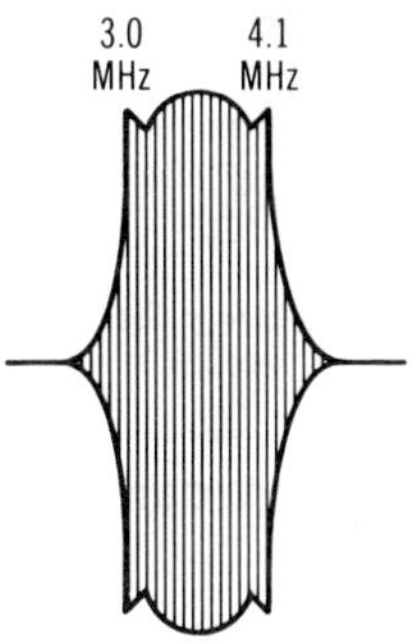

(D) Pattern at bandpass amplifier output.

modulation test setup.

NOTE 3–18

Check of Sweep-Generator Output Signal

Most sweep generators have a virtually flat output over any vhf channel. However, if a defect develops in the generator, its output may not be flat. To check this possibility, connect the output from the sweep generator directly to the input of the demodulator probe shown in Fig. 3-39. A flat-topped pattern should be displayed; if not, the sweep generator is in need of service. Note that two representative double-ended demodulator probe circuits are shown in Fig. 3-39. Either configuration is suitable. Most vhf sweep generators have double-ended (push-pull) output. In other words, both of the generator output leads operate above ground.

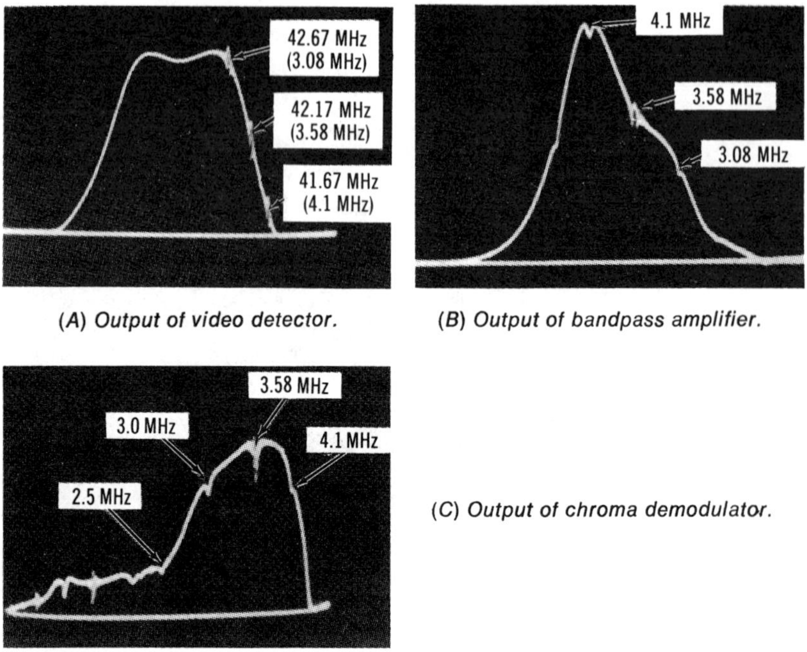

(*A*) *Output of video detector.*

(*B*) *Output of bandpass amplifier.*

(*C*) *Output of chroma demodulator.*

Fig. 3-37. Examples of demodulated VSM frequency response curves.

3–25. To Check the Input Impedance of a TV Tuner

Equipment: Vhf sweep generator, sample length of lead-in of the type normally used with the tuner.

Connections Required: With reference to Fig. 3-40, connect one end of a section of lead-in to the input terminals of the tuner. Connect the other end of the lead-in section to the output of a vhf sweep generator, and also to the input terminals of a double-ended demodulator probe.

Procedure: Adjust the tuner for normal operation, and operate the sweep generator on the same channel as the tuner. Use approximately 6 MHz deviation (sweep width).

Evaluation of Results: If the input impedance of the tv tuner is the same as the characteristic impedance of the lead-in, a flat-topped pattern will be displayed, as shown in Fig. 3-39. Impedance mismatch shows up as a slope or other irregularity in the top of the pattern. Each of the tuner channels should be checked in the same manner.

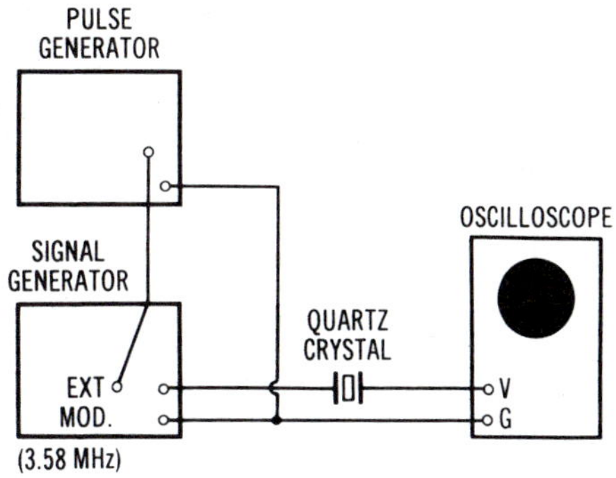

(A) Test setup.

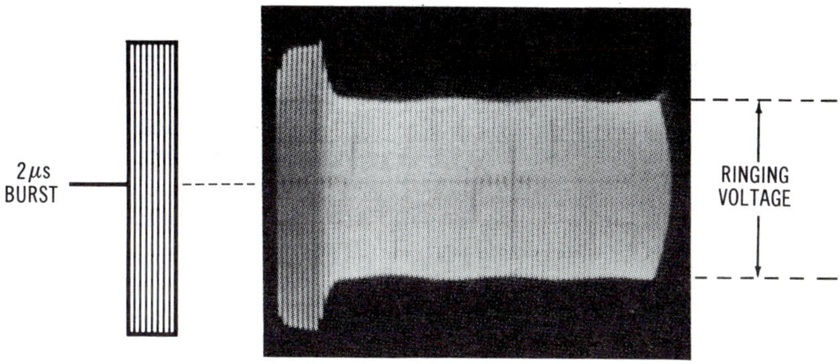

(B) 3.58-MHz ringing pattern (Q=8000 approximately).

Fig. 3-38. Pulse ringing test of color subcarrier oscillator crystal.

NOTE 3–19

Measurement of Tuner Input Impedance

In case that a tv tuner is mismatched on one or more channels with respect to its rated value of impedance, it is possible to measure the tuner impedance by a follow-up test procedure. With reference to the diagram in Fig. 3-40, the amount of slope and/or curvature in the scope pattern is noted; this pattern will then be duplicated. To do so, disconnect the tuner, and substitute various values of resistance. Thereby, the mismatch pattern can either be matched or approximated. If approximated, a precise match can be obtained by using various values of small capacitances and/or various small inductance values in series with resistance. Sometimes, a small value of capacitance will be needed in shunt to the resistance. Finally, the actual impedance of the tuner input can be accurately determined.

3–26. To Check the Impedance of an Antenna

Equipment: Vhf sweep generator.

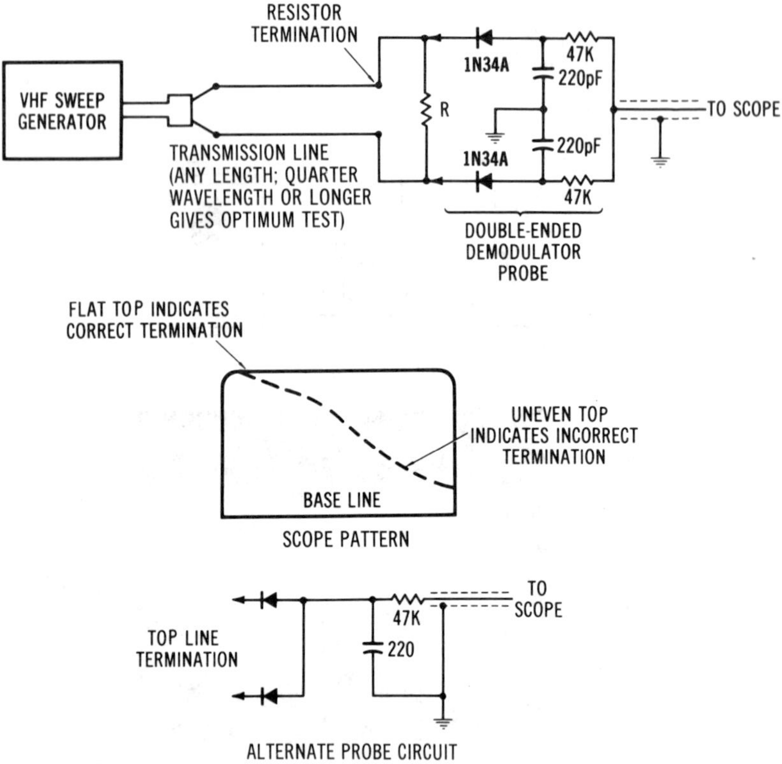

Fig. 3-39. Measurement of lead-in (transmission line) characteristic impedance.

Connections Required: Connect output from vhf sweep generator to lead-in, and also connect the lead-in to a double-ended demodulator probe, as shown in Fig. 3-41.

Procedure: Set sweep generator for 10-MHz deviation (sweep width) and observe the scope patterns as the generator out-

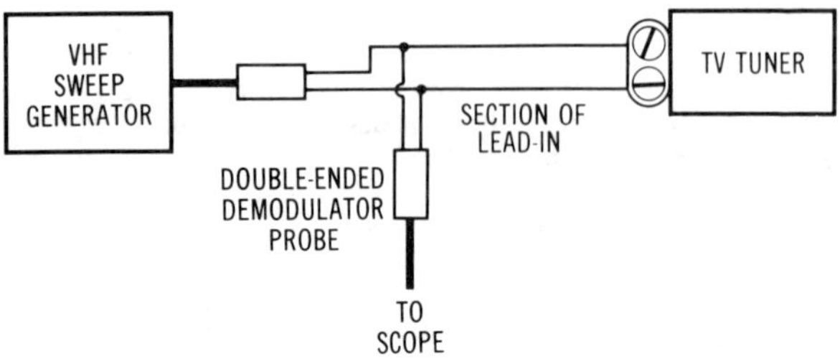

Fig. 3-40. Check of tv tuner input impedance.

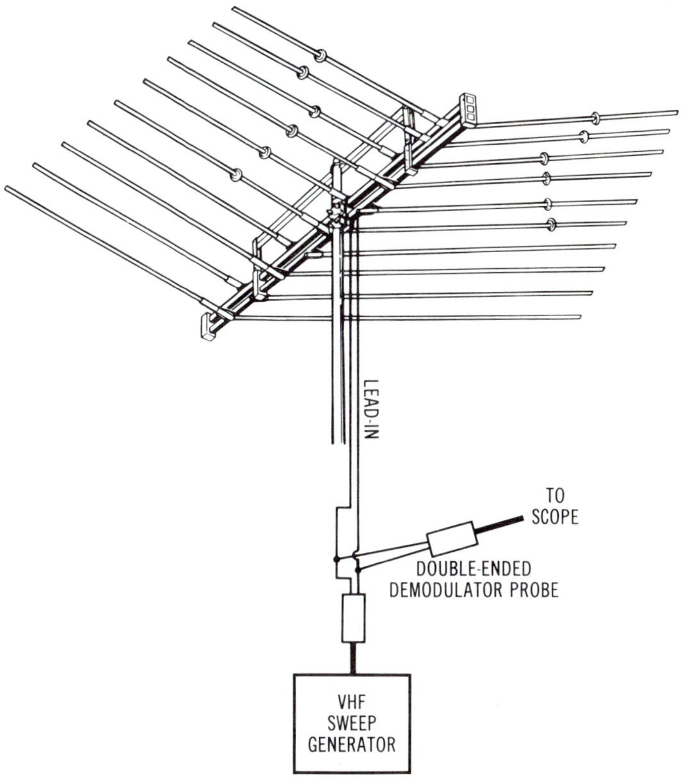

Fig. 3-41. Check of antenna impedance.

put frequency is switched progressively from channel 2 through 13.

Evaluation of Results: If the antenna impedance is the same as the characteristic impedance of the lead-in, a flat-topped pattern will be displayed, as shown in Fig. 3-39. Impedance mismatches show up as uphill or downhill slopes and/or curvatures in the tops of the patterns.

NOTE 3–20

Test of Coaxial Cable Installation

Although 300-ohm twin lead is the most widely used type of tv transmission line, some installations employ 75-ohm coaxial cable. The foregoing tests can be easily made in coaxial cable systems. However, since there is only one "hot" conductor in a cable circuit, a single-ended (conventional) demodulator probe is utilized. Note that single-ended vhf output is also used in the generator connections in this case; if a single-

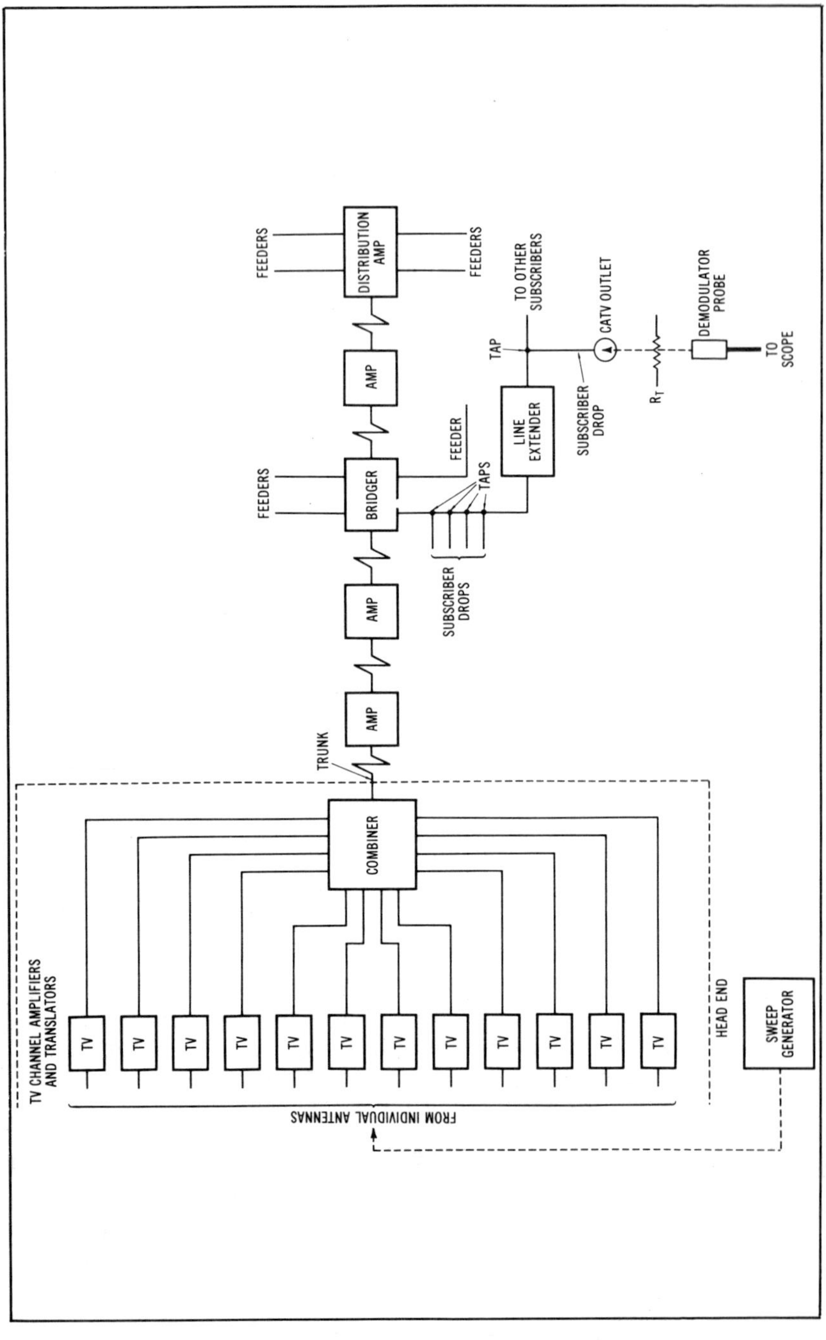

Fig. 3-42. Frequency response check of a catv system.

ended vhf output cable is not provided, one-half of the double-ended output cable may be used. In other words, the coaxial cable under test is driven from one of the "hot" output leads, and the shield of the coaxial cable is returned to the ground lead of the generator output cable. Note also that the foregoing tests are valid, whether the generator output cable matches or mismatches the twin lead or the coaxial cable under test.

3–27. To Make a Sweep Frequency Check of a CATV System

Equipment: Vhf sweep generator, terminating resistor for a catv outlet.

Connections Required: With reference to Fig. 3-42, connect the output of the sweep generator to the input of a chosen channel amplifier. Connect a suitable terminating resistor across the terminals of a chosen catv outlet. Connect a scope via a single-ended demodulator probe across the terminating resistor.

Procedure: Adjust the sweep generator to the center frequency of the channel under test, and use approximately 10-MHz sweep width. Observe the resulting pattern on the scope screen.

Evaluation of Results: In normal operation, a flat-topped pattern will be displayed, as was depicted in Fig. 3-39. Otherwise, a frequency compensator in an associated amplifier or line extender is in need of adjustment. These are called tilt controls, or compensators. (If the top of the pattern slopes uphill or downhill, it is said to be tilted.) *It is good practice to make an initial test at the most remote catv outlet. A flat response at this point verifies that the main portion of the system is operating properly. However, a comprehensive checkout requires that an outlet along each feeder from a bridger, line extender, or distribution amplifier be checked for frequency response.*

NOTE 3–21

Application of Wide-Band Sweep Generator

To speed up sweep-frequency checks of catv systems, a wide-band sweep generator may be used which sweep all 12 of the vhf channels. Similarly, wide-band generators that sweep channels 14 through 83 simultaneously may be used. Matv installations can also be checked with either conventional or with wide-band sweep generators. Note that low-level systems may require demodulator probes with suitable preamps.

3–28. To Measure the Distance to a Short or an Open in a Cable

Equipment: Pulse generator.

Connections Required: Connect output from pulse generator to input end of cable. Also connect input end of cable to vertical channel of scope, as shown in Fig. 3-43.

Procedure: Drive the cable with a comparatively narrow pulse, and observe the relative position of the reflected pulse in the scope pattern.

Evaluation of Results: A reflected pulse that has the same polarity as the initial pulse indicates that the cable is open-circuited at its far end. On the other hand, a reflected pulse that has opposite polarity from the initial pulse indicates that the cable is short-circuited at its far end. The elapsed time from the peak of the initial pulse to the peak of the reflected pulse is proportional to the distance from the generator to the cable defect. For example, suppose that the reflected pulse has opposite polarity to the initial pulse, and arrives after an elapsed time of 5.36 μs. This means that the cable is shorted at a distance corresponding to 2.18 μs (one-half the total travel time). If an air-dielectric cable were utilized, this would correspond to a distance of one-half mile (a pulse

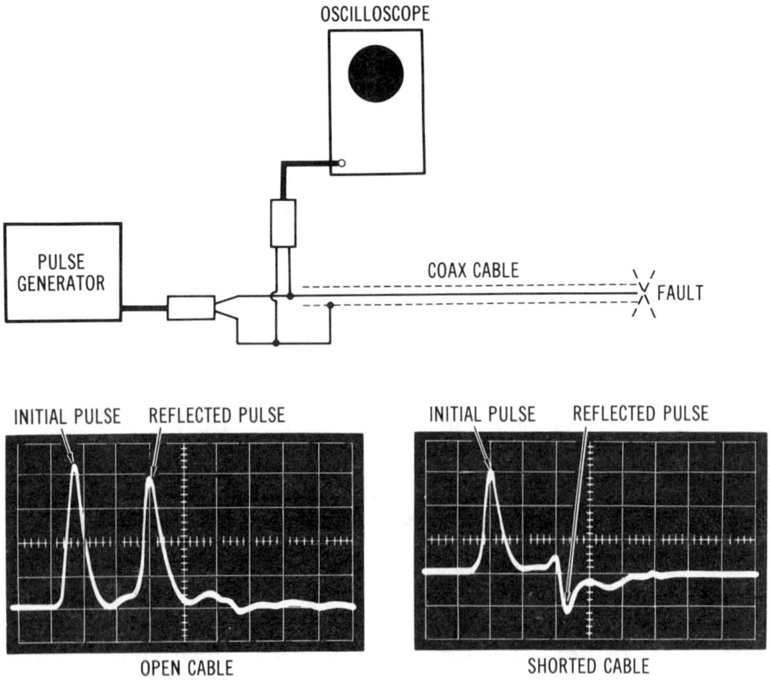

Fig. 3-43. Measurement of distance to a cable fault.

travel rate of 328 yards/μs). If a solid-dielectric cable is utilized, the pulse travel rate will be slower; thus, if the rated velocity constant of the cable is 0.8, the distance to the short-circuit will be 0.5/0.8 = 0.625 mile (5/8 mile).

3–29. To Check the Operation of a Delay Line

Equipment: Pulse generator.

Connections Required: Connect output from pulse generator to input of delay line. Connect A channel of dual-trace scope to the input of the delay line. Connect B channel of scope to output of delay line, as shown in Fig. 3-44.

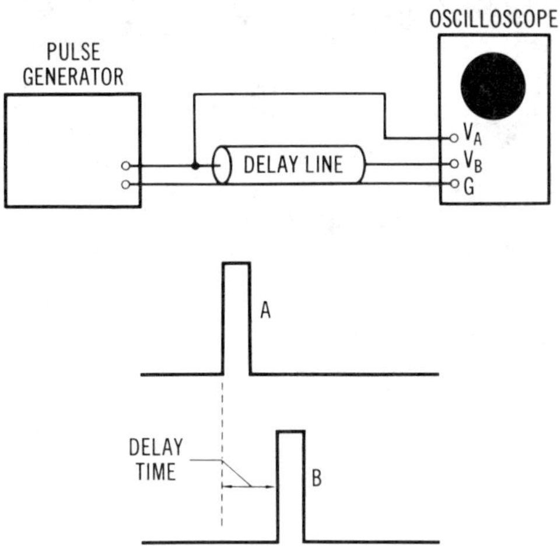

Fig. 3-44. Checking delay line action.

Procedure: Observe the displacement of the output pulse with respect to the input pulse.

Evaluation of Results: The displacement of the output pulse is proportional to the delay time. For example, if the sweep speed is 1 cm/μs, and the pulse displacement is 0.9 cm, the corresponding delay time is 0.9 μs. Note that pulse delay is always referenced to the 50 percent of maximum amplitude on the input pulse, and to the 50 percent of maximum amplitude on the output pulse. This method avoids possible confusion or error resulting from pulse rise times, and eliminates cornering characteristics from the calculation.

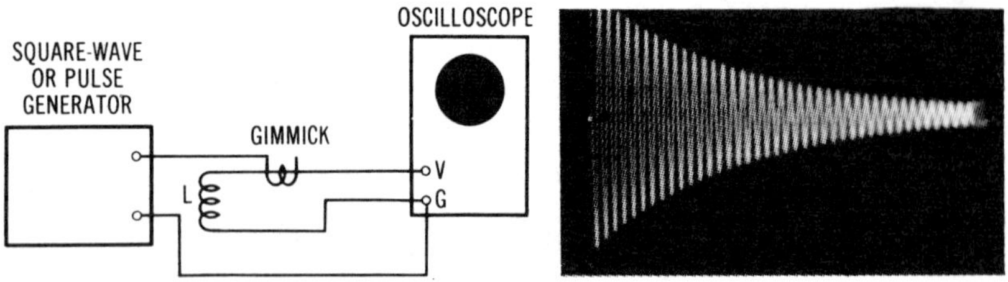

(*A*) *Test setup.*

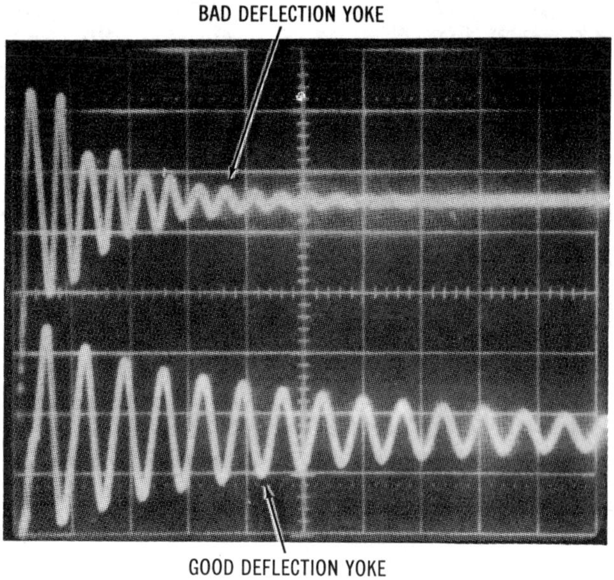

(*B*) *Testing a yoke.*

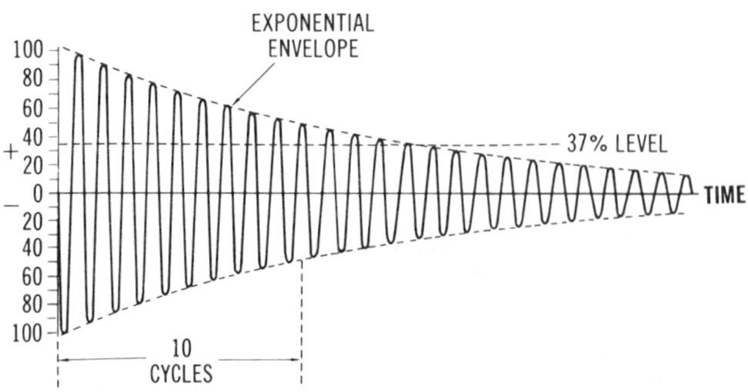

(*C*) *Calculating approximate Q value.*

Fig. 3-45. Ringing test for yoke or other inductor.

NOTE 3–22

Rise Time of Pulses on Lines

Pulses displayed in checking the operation of a delay line do not have zero rise time. In other words, the leading edges of the input and output pulses tend to slope. The output pulse may display somewhat more slope than the input pulse. In any case, precise measurement of delay time is made by observing the elapsed time from the 50 percent point on the leading edge of the input pulse to the 50 percent point on the leading edge of the output pulse.

3–30. To Make a Ringing Test of a Yoke (or Other Inductor)

Equipment: Square-wave or pulse generator that provides fast rise.

Connections Required: Connect equipment as shown in Fig. 3-45. (Use small capacitor instead of gimmick for a large coil.)

Procedure: Display the ringing waveform as illustrated.

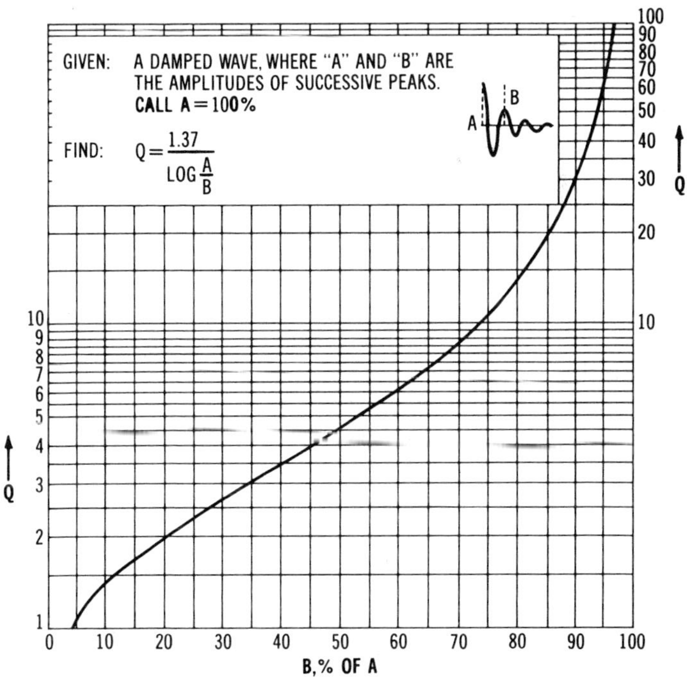

Fig. 3-46. Chart of Q values for a ringing system in terms of successive peak amplitudes.

Evaluation of Results: In the case of a yoke or flyback transformer, a comparative test is made; the pattern for a suspected yoke is compared with the pattern for a known good yoke. In other applications, the Q value of a coil is determined from its ringing pattern. Count the number of peaks (cycles) in the pattern from its 100 percent to its 37 percent amplitude point. Multiply this number of cycles by pi (3.14); this gives the approximate Q value of the coil at its natural resonant frequency. The ringing frequency is equal to the coil's natural resonant frequency, and is easily measurable with the calibrated time base in the scope. This natural resonant frequency results from the inductance and the distributed capacitance of the coil.

NOTE 3–23

Precise Measurement of Q Value

When it is desired to measure the Q value of a coil precisely, reference may be made to the chart in Fig. 3-46. It is based on the rate of decay of a ringing waveform. For example, suppose that each peak in a ringing waveform has 10 percent less amplitude than the preceding peak. In such a case, the chart shows that the Q value of the coil is 30. By way of comparison, the approximate method gives a value of 31.4 for Q in this example; thus, the approximate method is in error by less than 5 percent.

SECTION 4

Digital Logic Tests

4–1. To Check the Operation of an AND Gate

Equipment: Pulse generator or square-wave generator, battery.

Connections Required: Connect scope at output of AND gate. Connect output from generator in turn to each of the inputs of the AND gate. Finally, connect the output from the generator to all of the inputs of the AND gate. Use rated V_{cc} supply voltage.

Procedure: Apply a suitable output level from the generator, such as 2.5 volts peak. Observe the scope screen as the foregoing test connections are made. (See diagram in Fig. 4-1.)

Evaluation of Results: An output pulse should be displayed on the scope screen only when the generator is connected to all of the AND gate inputs. Otherwise, the gate is defective and should be replaced. (In a noisy environment, a gate may appear to be defective, although it is actually normal, unless the unused gate inputs are grounded to "kill" noise-voltage pickup.)

NOTE 4–1

NAND Gate Characteristics

A NAND gate is checked in the same manner as an AND gate. The only distinction is that a NAND gate inverts the polarity of the output pulse with respect to the input pulses. Note that an AND gate or a NAND gate has a minimum of two input terminals; however, these gates may have any maximum number of input terminals. For example, 8-input AND and NAND gates are often encountered (Fig. 4-2). However, the op-

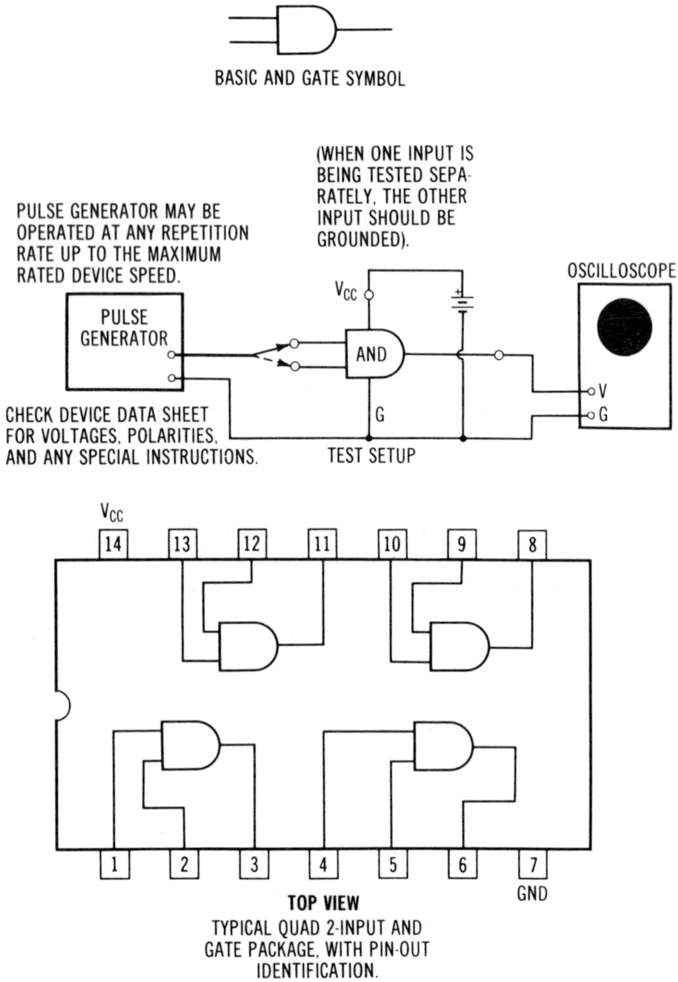

Fig. 4-1. Checking the operation

erating principle remains the same, regardless of the number of inputs. For example, if an AND gate has 8 inputs, an output pulse will normally be obtained only when all eight inputs are simultaneously driven logic-high. Always check the manufacturer's data for correct voltages and polarities. Note also that gates are generally fabricated with several gates included in a single IC package. Thus, a quad 2-input AND gate is an IC package that contains four AND gates, each with two inputs.

4–2. To Check AND Gate Action in a Scanner-Monitor Receiver

Equipment: None

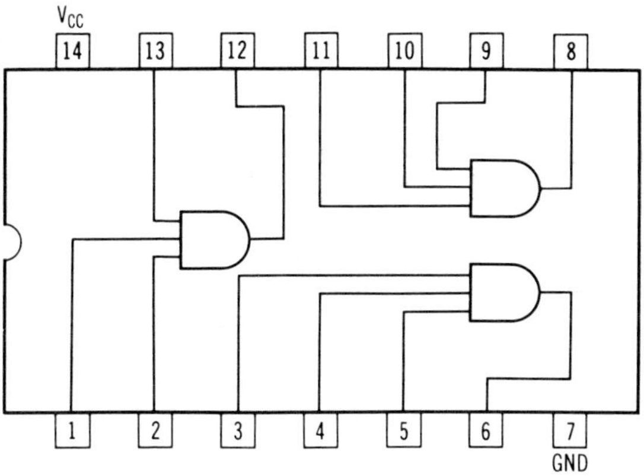

TRIPLE 3-INPUT AND GATE
PACKAGE

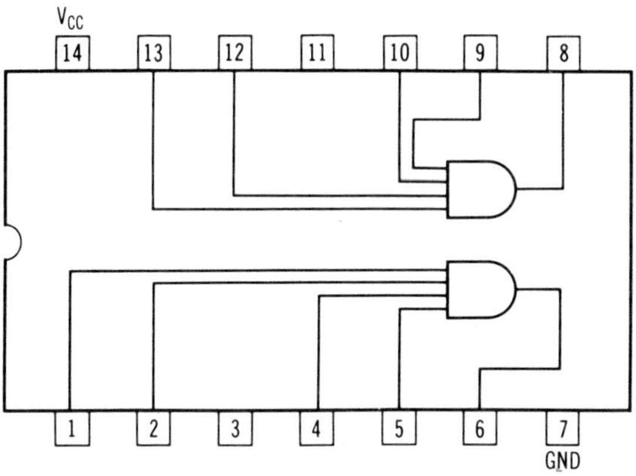

DUAL 4-INPUT AND GATE
PACKAGE

of an AND gate.

Connections Required: With reference to Fig. 4-3, connect scope via lo-C probe at the clock output, FF1 Q output, FF1 $\overline{Q}$ output, FF2 Q output, and FF2 $\overline{Q}$ output, in turn.

Procedure: Observe the waveform displayed (or absence of waveform) at each of the test points.

Evaluation of Results: A square-wave train (clock pulses) are normally present at the clock output; half-frequency square waves in opposite polarities are normally present at the FF1

Q and the FF1 $\overline{Q}$ outputs; quarter-frequency square waves in opposite polarities are normally present at the FF2 Q and the FF2 $\overline{Q}$ outputs. Incorrect output, or absence of output, at a

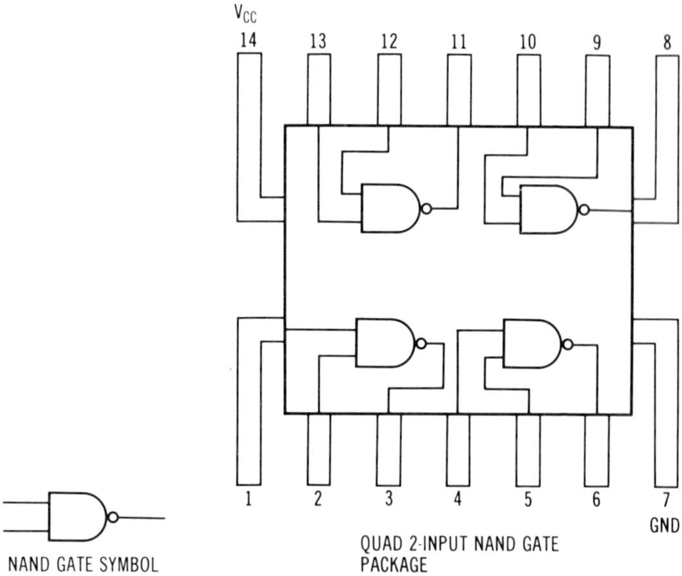

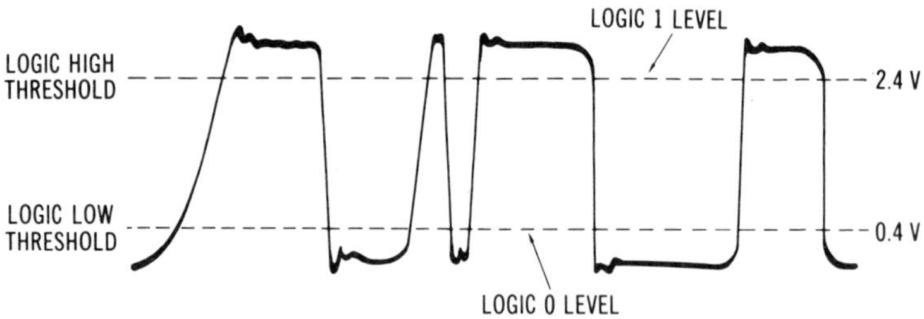

Fig. 4-2. Typical digital waveform in TTL circuitry, showing High (1) and Low (2) thresholds. (CMOS circuitry employs supply voltages in the range from 3 to 18 volts; check manufacturer's data sheet).

test terminal indicates a fault in the associated device. The array of square waves depicted in the diagram comprises the specified *timing diagram* for the digital system.

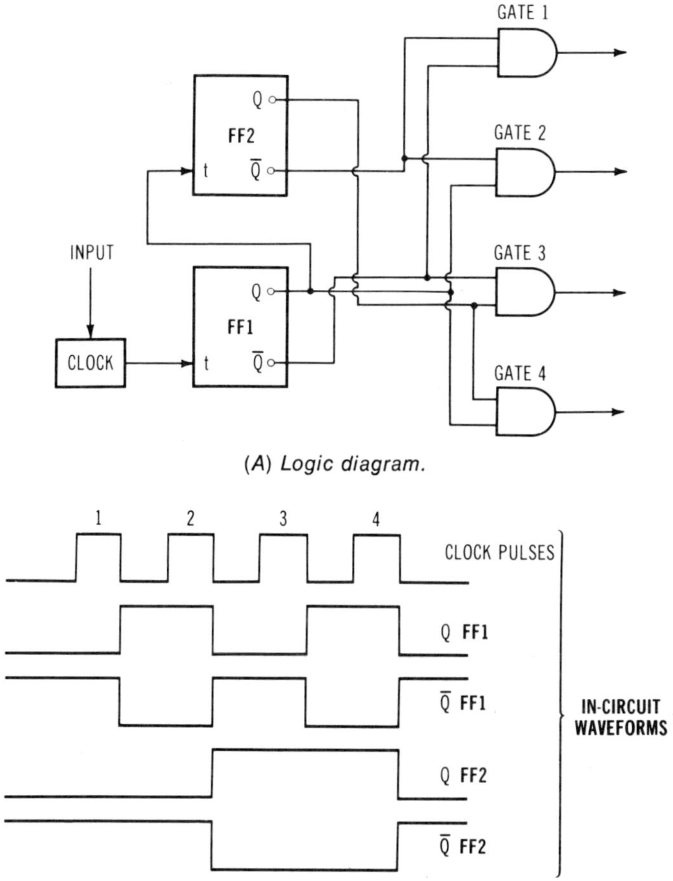

(*A*) *Logic diagram.*

(*B*) *Normal operating waveforms.*

Fig. 4-3. Simplified block diagram for a 4-channel digital-logic section in a scanner-monitor radio receiver.

NOTE 4–2

Digital Clock Function

A *clock* is a free-running multivibrator which functions to synchronize circuit actions throughout a digital system. A *flip-flop* (FF) is a bistable multivibrator; when the FF is "on," its Q output is logic-high—when "off," its $\bar{Q}$ (NOT Q) is logic-low. The logic-high state is symbolized by 1, and the logic-low state is symbolized by 0. When the Q output of a FF is 1, its $\bar{Q}$ output is 0, and vice versa. The Q and $\bar{Q}$ outputs are said to be *complementary*. Each AND gate has a 0 output unless both of its inputs are simultaneously 1. If both AND gate inputs are 0, its output is 0; if the AND gate inputs are 1,0 or 0,1, its output is 0. *Note that the timing diagram in Fig. 4-3 represents a positive operating level, with positive*

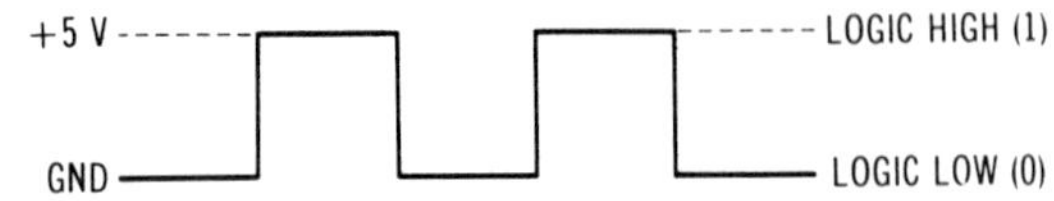

(A) *Positive operating level, with positive logic.*

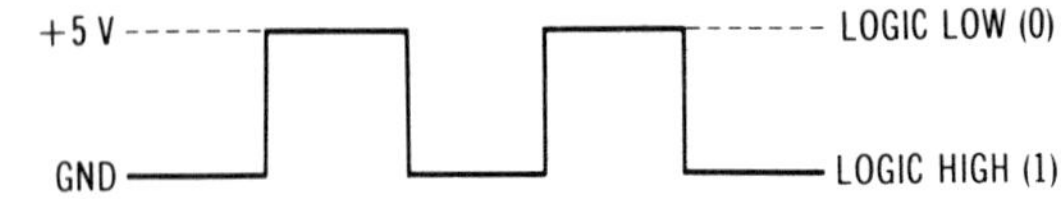

(B) *Positive operating level, with negative logic.*

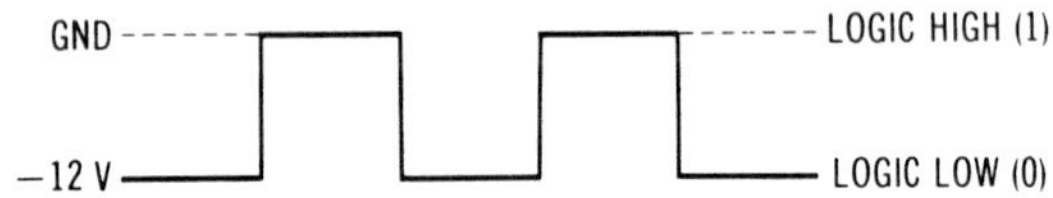

(C) *Negative operating level, with negative logic.*

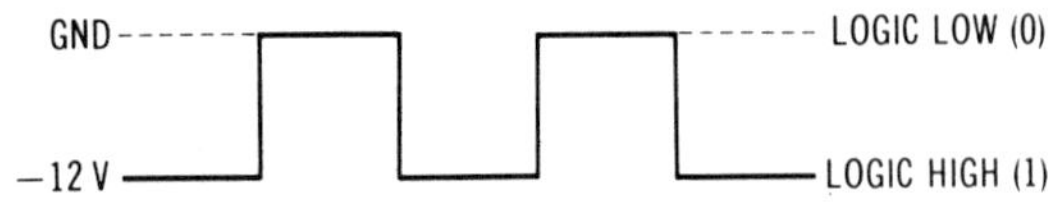

(D) *Negative operating level, with positive logic.*

Fig. 4-4. Basic example of positive and negative operating levels, with positive and negative logic conventions.

logic. However, the technician will encounter instances of positive operating level with negative logic, of negative operating level with negative logic, and of negative operating level with positive logic, as exemplified in the diagram of Fig. 4-4.

4–3. To Check the Operation of an OR Gate

Equipment: Pulse generator or square-wave generator, and battery.

Connections Required: Connect scope at output of OR gate. Connect output from generator in turn to each of the inputs of the OR gate. Finally, connect the output from the generator to all of the inputs of the OR gate. Use rated V_{cc} supply voltage.

Procedure: Apply a suitable output level from the generator, such as 2.5 volts peak. Observe the scope screen as the foregoing test connections are made. (See Fig. 4-5.)

Evaluation of Results: An output pulse should be displayed on the scope screen whenever the generator output is applied

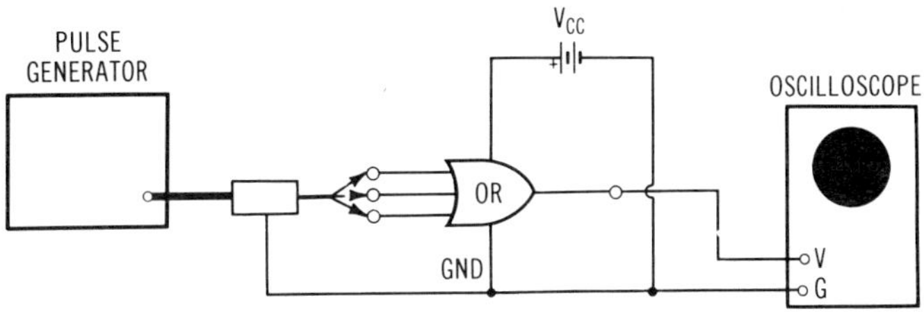

(*A*) *Test setup.*

(*B*) *Basic OR gate symbol.*

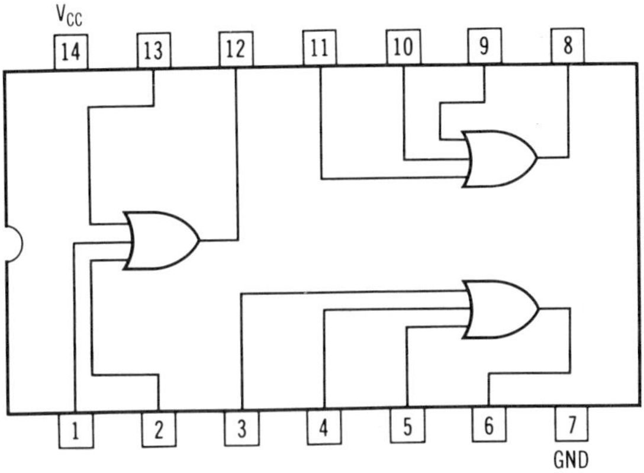

(*C*) *Typical triple 3-input OR gate package.*

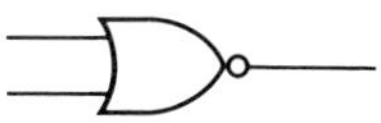

(*D*) *NOR gate symbol.*

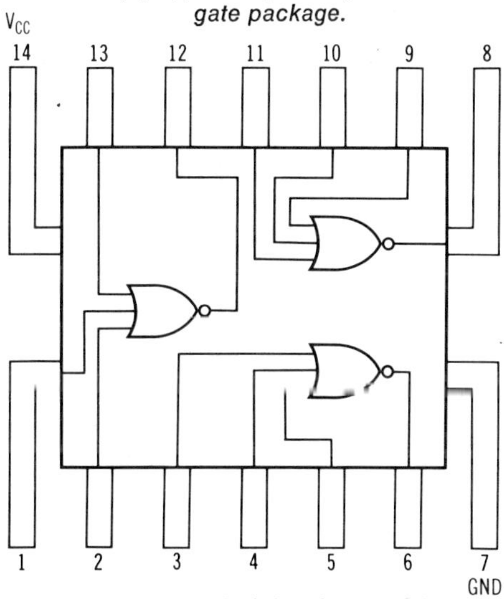

(*E*) *Typical triple 3-input NOR gate package.*

Fig. 4-5. Checking the operation of an OR gate.

to one or more of the OR gate inputs. Otherwise, the gate is defective and should be replaced.

NOTE 4–3
NOR Gate Characteristics

A NOR gate is checked in the same manner as an OR gate. The only distinction is that a NOR gate inverts the polarity of the output pulse with respect to the input pulse(s). Note that an OR gate or a NOR gate has a minimum of two input terminals; however, these gates may have any maximum number of input terminals. The operating principle remains the same, regardless of the number of inputs. For example, if an OR gate has five inputs, an output pulse will normally be obtained whenever any input terminal, or any combination of input terminals, is driven logic-high. Always check the manufacturer's data for correct voltages and polarities.

4–4. To Check the Operation of an XOR Gate

Equipment: Pulse generator or square-wave generator, and battery.

Connections Required: Connect scope at output of XOR gate. Connect output from generator in turn to each of the inputs of the XOR gate. Finally, connect the output from the generator to both of the inputs of the XOR gate (Fig. 4-6). Use rated V_{cc} supply voltage.

Procedure: Apply a suitable output level from the generator. Observe the scope screen as the foregoing test connections are made.

Evaluation of Results: An output pulse should be displayed on the scope screen whenever the generator output is applied to one of the XOR gate inputs. On the other hand, no output pulse should be displayed when the generator output is applied simultaneously to both of the XOR gate inputs. Failure to respond normally to these tests indicates that the XOR gate is defective.

NOTE 4–4
$\overline{\text{XOR}}$ Gate Characteristics

An EXCLUSIVE OR (XOR) gate has two inputs; it produces a logic-high output only when its inputs are driven to opposite levels. If its inputs are 1,1 or 0,0, the XOR gate produces no output (is at a logic-low level). An EXCLUSIVE NOR ($\overline{\text{XOR}}$) gate is checked in the same manner as an XOR gate. The only distinction is that an $\overline{\text{XOR}}$ gate inverts the output that would be obtained with an XOR gate. In other words, an $\overline{\text{XOR}}$ gate produces a logic-low output only when its inputs are driven

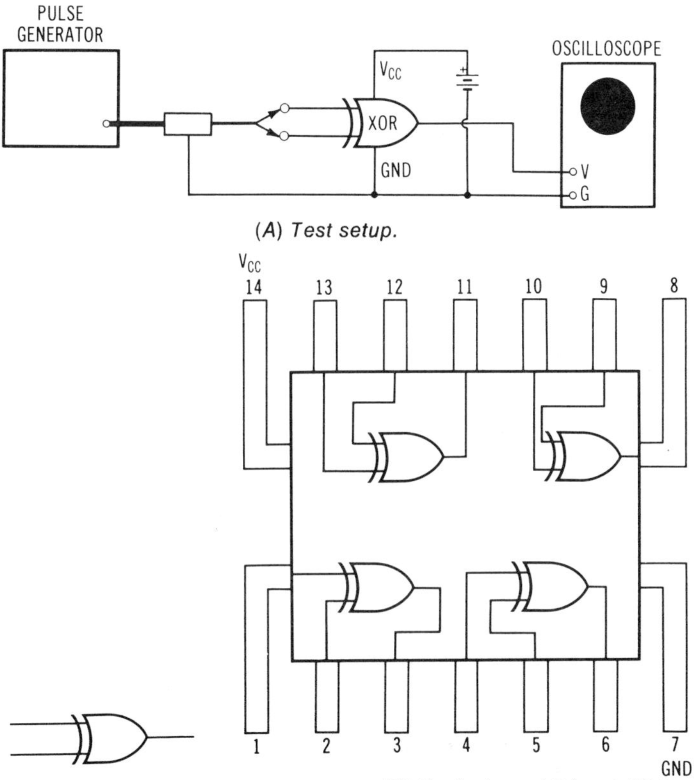

(A) Test setup.

(B) Basic XOR gate symbol.

(C) Typical quad 2-input XOR gate package.

Fig. 4-6. Checking the operation of an XOR gate.

to opposite levels. If the $\overline{XOR}$ inputs are 1,1 or 0,0, the gate output is at a logic-high level. The normal responses of AND, OR, NAND, NOR, XOR, and $\overline{XOR}$ (XNOR) gates are summarized to good advantage by means of truth tables as shown in Fig. 4-7.

4-5. To Check the Operation of an AND-OR-INVERT Gate

Equipment: Pulse or square-wave generator, and battery.

Connections Required: With reference to Fig. 4-8, connect output terminal of device to vertical-input channel of scope. Connect output from generator to AND gate inputs, as explained next. Connect rated V_{cc} supply voltage from V_{cc} terminal to ground.

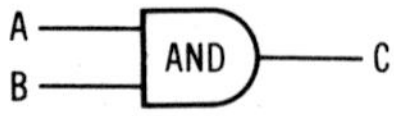

A	B	C
1	0	0
0	1	0
1	1	1
0	0	0

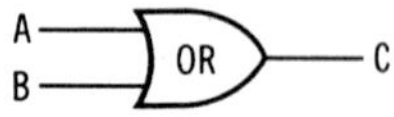

A	B	C
1	0	1
0	1	1
1	1	1
0	0	0

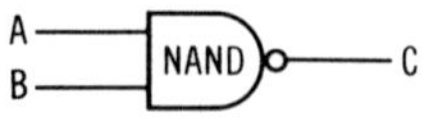

A	B	C
1	0	1
0	1	1
1	1	0
0	0	1

A	B	C
1	0	0
0	1	0
1	1	0
0	0	1

A	B	C
1	0	1
0	1	1
1	1	0
0	0	0

A	B	C
1	0	0
0	1	0
1	1	1
0	0	1

Fig. 4-7. Truth tables summarize normal responses.

Procedure: Use an appropriate output level from the square-wave generator at any repetition rate within the specified range of the device. Apply the test signal in turn to each individual AND-gate input terminal, and observe scope screen. Then, apply the test signal in turn to each set of AND-gate input terminals, and observe scope screen.

Evaluation of Results: In normal operation, no waveform will be displayed on the scope screen unless the test signal is applied simultaneously to a set of AND-gate input terminals. Otherwise, the device is defective and should be replaced.

4–6. To Check the Operation of a Full Adder

Equipment: Pulse generator or square-wave generator, and battery.

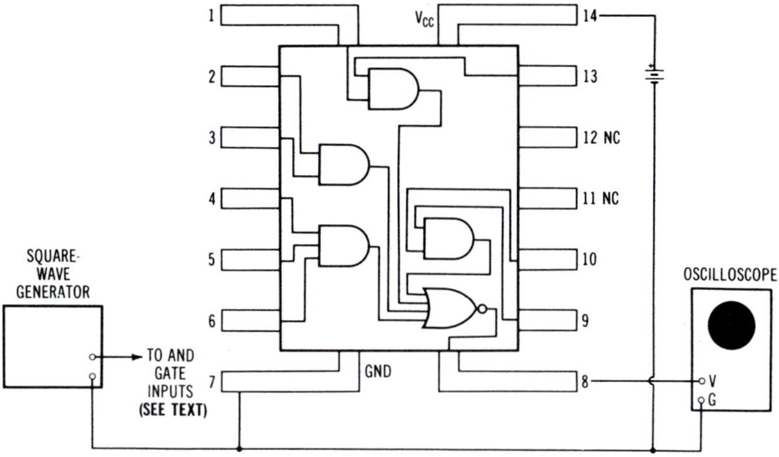

(A) Typical package pin-out and test setup.

(B) Basic four-wide 2-2-2-3 input AND-OR-INVERT gate symbol.

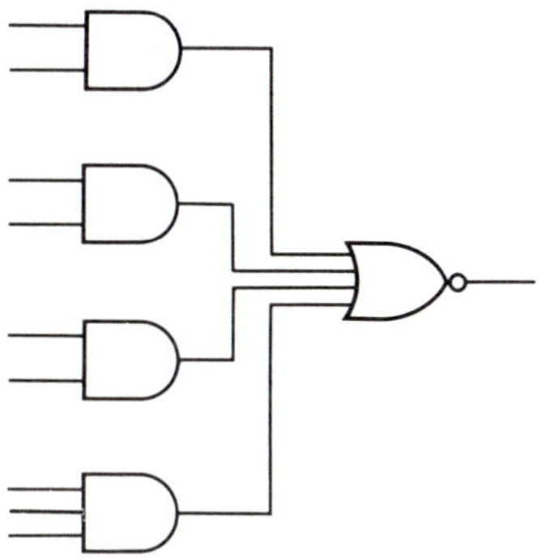

Fig. 4-8. Test of AND-OR-INVERT gate operation.

Connections Required: As shown in Fig. 4-9, connect a dual-trace scope at the Carry and Sum outputs of the full adder. Connect output from generator to one, two, or three inputs of the full adder, as successively listed in the truth table. Use rated V_{cc} supply voltage.

Procedure: Apply a suitable output level from the generator. Observe the scope screen as the foregoing test connections are made.

Evaluation of Results: An output pulse is normally displayed on channel A, on channel B, or both, when corresponding A, B, and C′ pulses are applied to the full adder. Failure to respond in accordance with the truth table indicates that the full adder is defective.

NOTE 4–5

Full Adder Operation

A full adder consists of two half adders with an OR gate. A half adder (Fig. 4-10) consists of an XOR gate and an AND gate (or equivalent

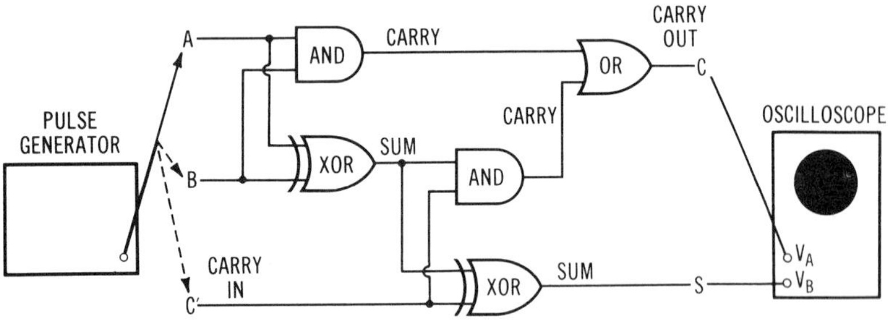

(*A*) *Test setup.*

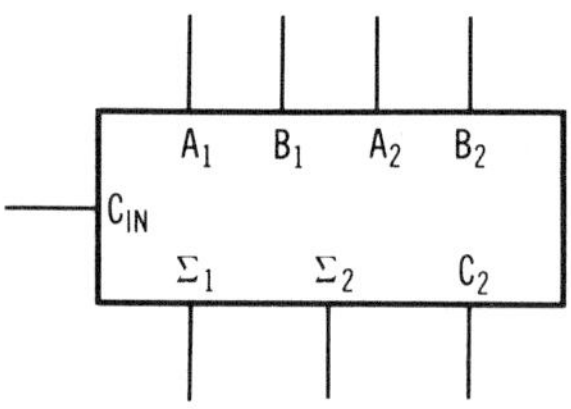

(*B*) *Basic full adder symbol.*

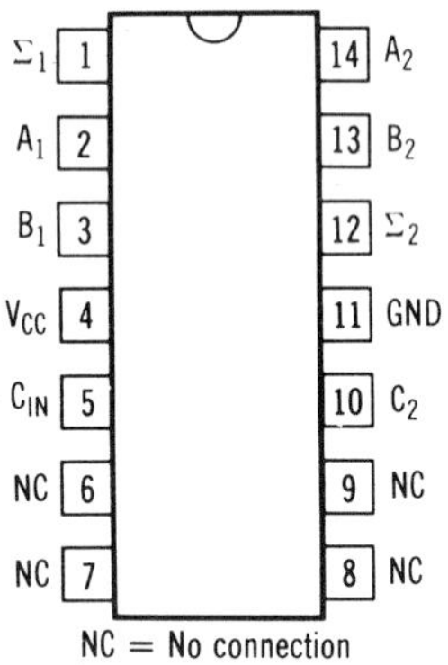

(*C*) *Typical package pin-out.*

A	B	C′	S	C
0	0	0	0	0
0	0	1	1	0
0	1	0	1	0
0	1	1	0	1
1	0	0	1	0
1	0	1	0	1
1	1	0	0	1
1	1	1	1	1

(*D*) *Truth table.*

Fig. 4-9. Check of full adder operation.

arrangement). A half adder has two inputs and a sum (S or Σ) output and a carry (C) output. A full adder has two inputs and a carry-in (C′) input, with a sum output and a carry-out output. The inputs are customarily designated A and B; pulses applied to the inputs are logic-high or logic-low levels and are called binary digits (bits). The foregoing adders are two-bit devices, since they have two inputs. Observe that the truth table for an adder is basically a statement of the rules for binary addition.

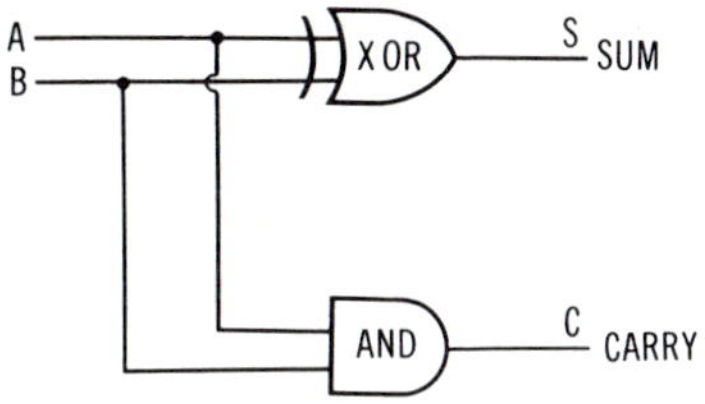

(*A*) *Gate arrangement.*

A	B	S	C
0	0	0	0
0	1	1	0
1	0	1	0
1	1	0	1

(*B*) *Truth table.*

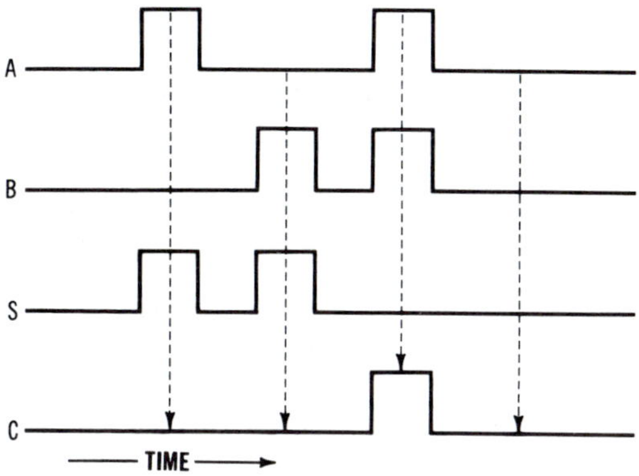

(*C*) *Operating waveforms.*

Fig. 4-10. Half adder.

4-7. To Check the Operation of a 4-Bit Latch

Equipment: Two square-wave generators, and battery.

Connections Required: As shown in Fig. 4-11, connect the output from one square-wave generator to the clock input of a selected latch, and connect the output from the other square-wave generator to the data input of the latch. Connect the A channel of a dual-trace scope to the Q output of the latch, and connect the data input of the latch to the B channel of the scope. If the square-wave generators have sync facilities, lock generator 2 from the output of generator 1. Use rated V_{cc} supply voltage.

Procedure: Operate the clock generator at two or three times the repetition rate of the data generator. Observe the Q output waveform with respect to the data input waveform.

Evaluation of Results: As indicated by the truth table, the Q output waveform normally reproduces the data input waveform, but lags by one-half of a clock cycle. Thus, if the clock gener-

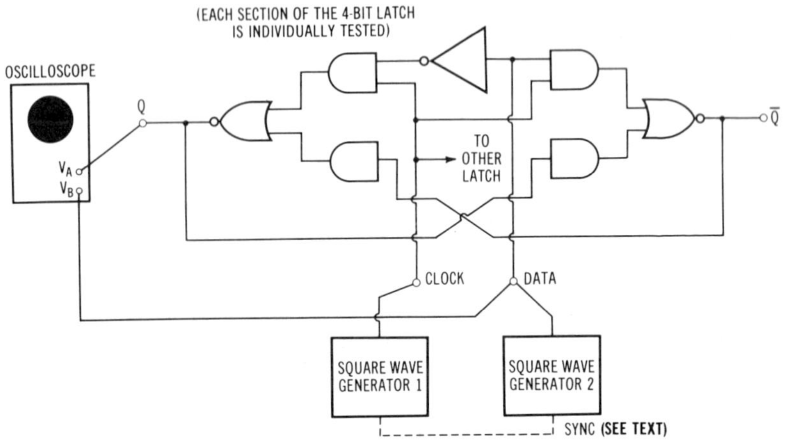

(A) Test setup.

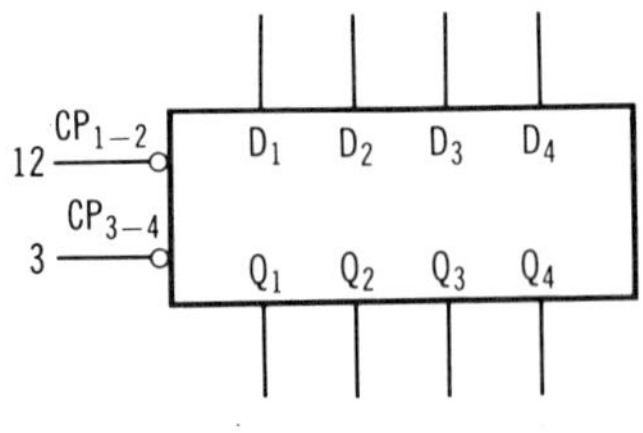

(B) Basic 4-bit latch symbol.

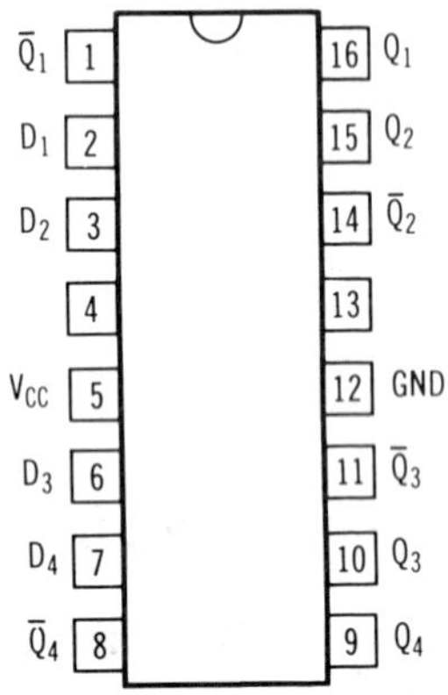

Positive logic: See truth table.
NC = No internal connection.

(C) Typical dual 4-bit latch package.

t_n	t_n+1
D	Q
1	1
0	0

NOTES:
t_n = Bit time before clock negative-going transition.
$t_n + 1$ = Bit time after clock negative-going transition.

(D) Truth table.

Fig. 4-11. Check of 4-bit latch operation.

ator is operating at twice the rate of the data generator, and the generators are synchronized, the Q output waveform normally lags the data input waveform by one-quarter cycle. Otherwise, the latch is defective and should be replaced.

NOTE 4–6

Latch Characteristics

A latch is basically a bistable multivibrator which is generally used as temporary storage for bits between processing units and input/output or indicator units. A high or a low state applied at the D input will be transferred to the Q output when the clock goes logic-high; thus, the Q output lags the D input waveform. This lag represents the temporary storage time. Note that the Q output from the latch normally has opposite polarity from the Q output (complements the Q output).

4–8. To Check the Operation of a JK Master/Slave Flip-Flop

Equipment: Two square-wave generators, and battery.

Connections Required: Connect the J and K inputs of the FF together and to the output from the square-wave generator 1 (Fig. 4-12). Connect the CP input to the output from square-wave generator 2. Connect the R_D input to a logic-high source (battery). Connect the Q and $\overline{Q}$ FF outputs respectively to the A and B channels of a dual-trace scope. Use rated V_{cc} supply voltage.

Procedure: Operate square-wave generator 2 at a higher repetition rate than square-wave generator 1. Use suitable output levels, per device data sheet. Observe the scope screen. Then connect the R_D input to ground, instead of the logic-high source, and again observe the scope screen.

Evaluation of Results: Since the J and K inputs are tied together, the flip-flop will normally *toggle;* when J and K are 0, Q remains in its previous (clock pulse) state; but when J and K go to 1, Q will normally change state. $\overline{Q}$ will normally complement Q's state. Thus, square-wave outputs are normally observed from the FF. On the other hand, when R_D is held at 0, Q normally remains at 0 and $\overline{Q}$ remains at 1. Otherwise, the FF is defective and should be replaced.

NOTE 4–7

JK Flip-Flop Functions

The JK master/slave flip-flop is in extremely wide use. It is a *clocked* device; inputs to the master section are controlled by the clock pulse. The clock pulse also controls the circuitry which connects the master section to the slave section. The normal sequence of operation is as follows: (1) Clock pulse isolates slave from master; (2) 1's or 0's are entered from J and K inputs to the master section; (3) Clock pulse disables J and K inputs; (4) Clock pulse also transfers master state to slave section. As shown by the truth table, the FF outputs will alternate (toggle) when the J and K inputs are tied together and driven by a square-wave voltage. The R_D input is ordinarily logic-high; however, the R_D input may be driven logic-low whenever desired, whereupon the Q output will be forced logic-low ($\overline{Q}$ forced logic-high). This forcing or clearing action of the R_D input can be employed at any time, and is independent of the clock action.

4–9. To Check the Operation of a 1-of-10 Decoder/Driver

Equipment: Pulse or square-wave generator, and battery.

Connections Required: With reference to Fig. 4-13, connect output from generator to specified inputs of decoder-driver. Connect specified output from decoder/driver to vertical-input channel of scope. Use rated V_{cc} supply voltage.

Procedure: First, with no signal input to any four of the inputs on the decoder/driver, observe the scope response at the various outputs. Next, with signal input at P_A only, observe the scope response at the various outputs. Continue with the sig-

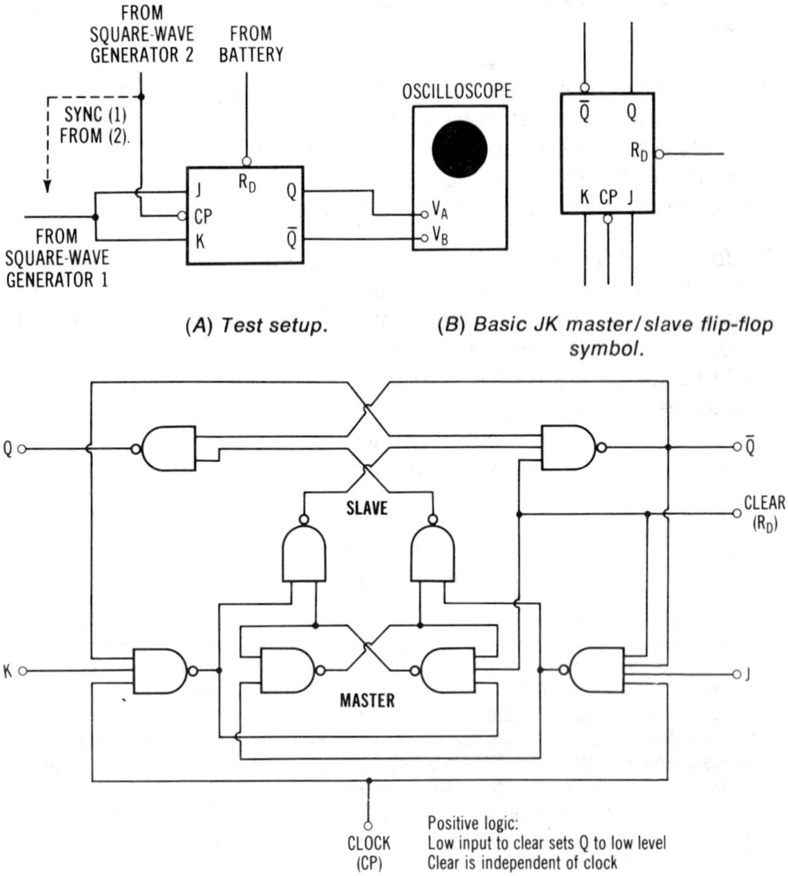

(*A*) *Test setup.* (*B*) *Basic JK master/slave flip-flop symbol.*

(*C*) *Basic logic diagram.*

Fig. 4-12. Check of JK

nal input combinations listed in the truth table, observing the scope response in each step.

Evaluation of Results: In normal operation, the scope will display a pulse or square-wave output only at the decoder/driver output terminals that correspond to the input signal conditions specified in the truth table. Otherwise, the decoder/driver is defective and should be replaced.

NOTE 4–8

Decoder-Driver Characteristics

The 1-of-10 decoder/driver exemplified in Fig. 4-13 is designed to accept binary-coded-decimal (bcd) inputs and to provide corresponding output states to drive 10-digit incandescent readout displays. All outputs normally remain off for all invalid input conditions. Note that the output gates in the decoder/driver may be provided as open-collector devices;

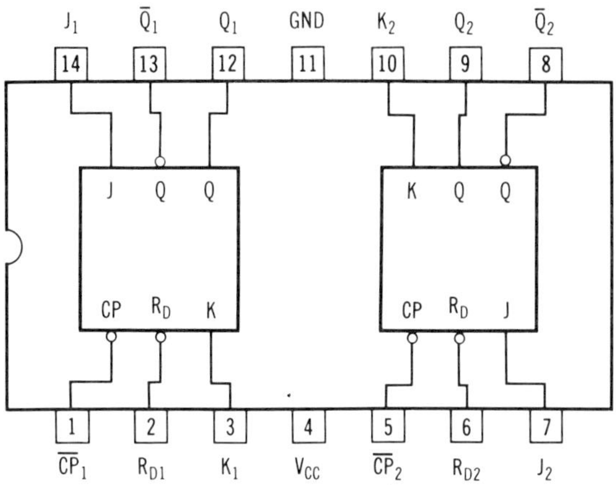

(D) Typical package pin-out.

t_n		$t_n + 1$
J	**K**	**Q**
0	0	Q_n
0	1	0
1	0	1
1	1	$\bar{Q}_n$

NOTES:
t_n = Bit time before clock pulse.
$t_n + 1$ = Bit time after clock pulse.

(E) Truth table.

flip-flop operation.

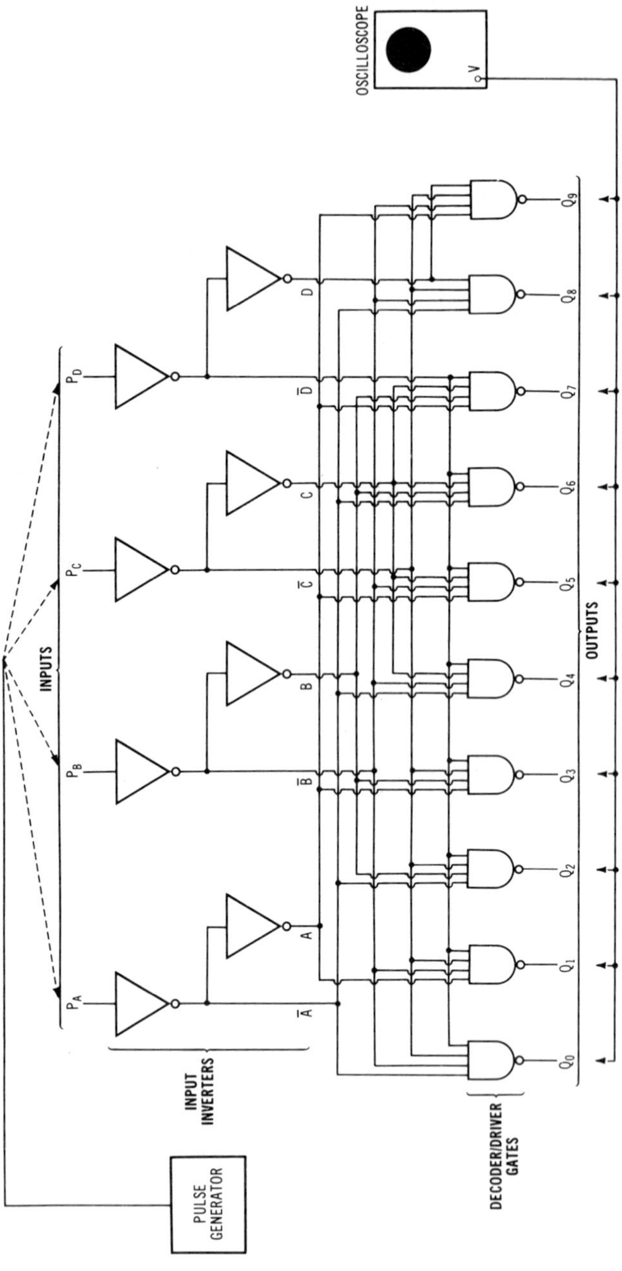

(A) Test setup.

Fig. 4-13. Check of 1-of-10

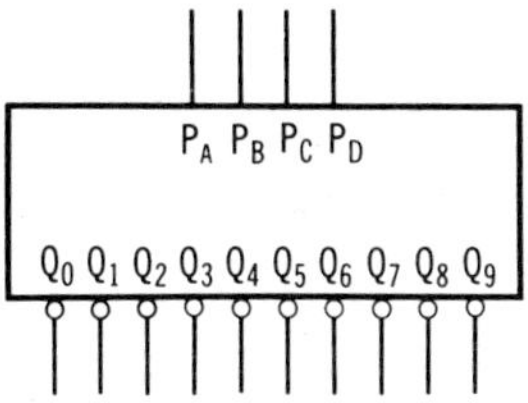

(B) Basic 1-of-10 decoder/driver symbol.

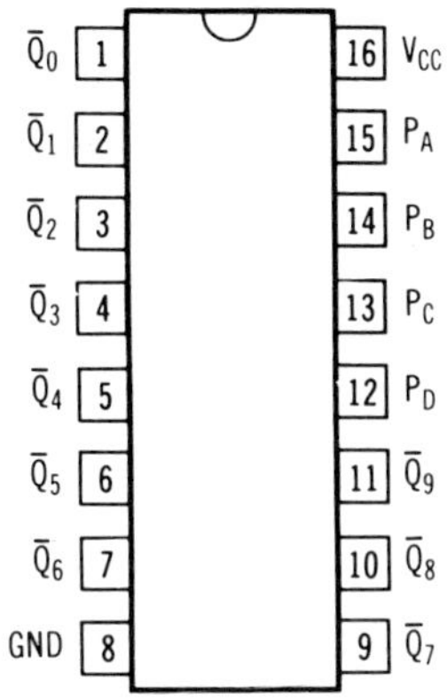

(C) Typical package pin-out.

INPUTS				OUTPUTS									
P_D	P_C	P_B	P_A	$\bar{Q}_0$	$\bar{Q}_1$	$\bar{Q}_2$	$\bar{Q}_3$	$\bar{Q}_4$	$\bar{Q}_5$	$\bar{Q}_6$	$\bar{Q}_7$	$\bar{Q}_8$	$\bar{Q}_9$
0	0	0	0	0	1	1	1	1	1	1	1	1	1
0	0	0	1	1	0	1	1	1	1	1	1	1	1
0	0	1	0	1	1	0	1	1	1	1	1	1	1
0	0	1	1	1	1	1	0	1	1	1	1	1	1
0	1	0	0	1	1	1	1	0	1	1	1	1	1
0	1	0	1	1	1	1	1	1	0	1	1	1	1
0	1	1	0	1	1	1	1	1	1	0	1	1	1
0	1	1	1	1	1	1	1	1	1	1	0	1	1
1	0	0	0	1	1	1	1	1	1	1	1	0	1
1	0	0	1	1	1	1	1	1	1	1	1	1	0
1	0	1	0	1	1	1	1	1	1	1	1	1	1
1	0	1	1	1	1	1	1	1	1	1	1	1	1
1	1	0	0	1	1	1	1	1	1	1	1	1	1
1	1	0	1	1	1	1	1	1	1	1	1	1	1
1	1	1	0	1	1	1	1	1	1	1	1	1	1
1	1	1	1	1	1	1	1	1	1	1	1	1	1

(D) Truth table.

decoder/driver operation.

in such a case, the device data sheet will include this information. Also, the test setup will require a load resistor of rated value at each output terminal. *Although a 1-of-10 decoder/driver operates with respect to binary-coded decimal notation, this is of no consequence in test procedures. In other words, the test procedure is concerned only with a determination of whether the device responds in accordance with its associated truth table.*

4–10. To Check the Operation of a Schmitt Trigger

Equipment: Sine-wave generator, and battery.

Connections Required: As shown in Fig. 4-14, tie all of the AND-gate inputs together and to the output from a sine-wave gen-

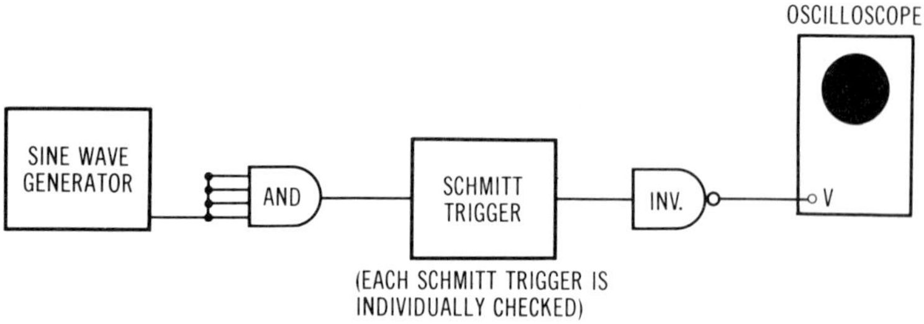

(A) Test setup.

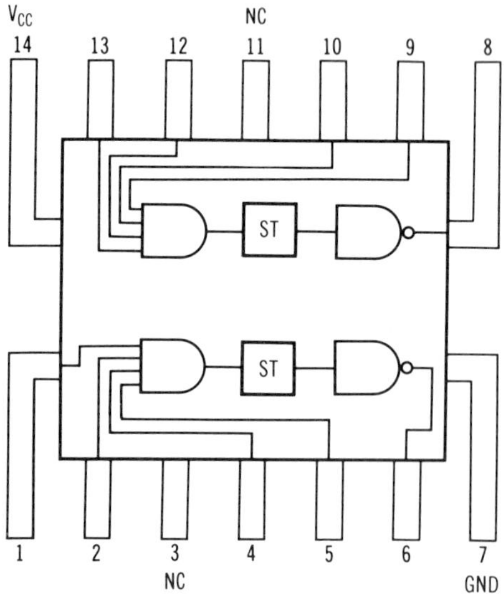

(B) Typical dual Schmitt trigger package pin-out.

Fig. 4-14. Checking operation of a Schmitt trigger.

erator. Connect output from Schmitt trigger to vertical-input channel of scope. Use rated V_{cc} supply voltage.

Procedure: Drive the Schmitt trigger with a suitable peak-to-peak ac voltage, according to the device data-sheet rating. Observe the scope screen.

Evaluation of Results: In normal operation, a square wave will be displayed on the scope screen; the square wave should have the same repetition rate as the sine-wave frequency. In other words, a Schmitt trigger normally operates as a waveform squaring circuit. For a comprehensive test, check the AND-gate inputs individually—there is normally zero output from the Schmitt trigger unless all of the AND-gate inputs are simultaneously driven logic-high, or logic-low.

4–11. To Check the Operation of a True/Complement Zero/One Element

Equipment: Square-wave generator, and battery.

Connections Required: Connect output from square-wave generator to data inputs of true/complement zero/one element, as shown in Fig. 4-15. Connect outputs from device to vertical-input channel of scope. Connect control inputs in turn to a logic-high source (battery) or to a logic-low source (ground). Use rated V_{cc} voltage.

Procedure: With both control inputs logic-high, observe the A_1-Y_1 response on the scope screen; continue through A_4-Y_4. Repeat procedure with both control inputs logic-low, and with one control input logic-high while the other control input is logic-low.

Evaluation of Results: As indicated in the truth table, the square-wave output normally duplicates the square-wave input on each channel when B is logic-high and C is logic-low. However, the square-wave output is normally inverted (complemented) when B and C are both logic-low. If B and C are both logic-high, all outputs are normally logic-low, and there is no square-wave passage. Or, when B is logic-high and C is logic-low, all outputs are normally logic-high, and there is no square-wave passage. Otherwise, the device is defective and should be replaced.

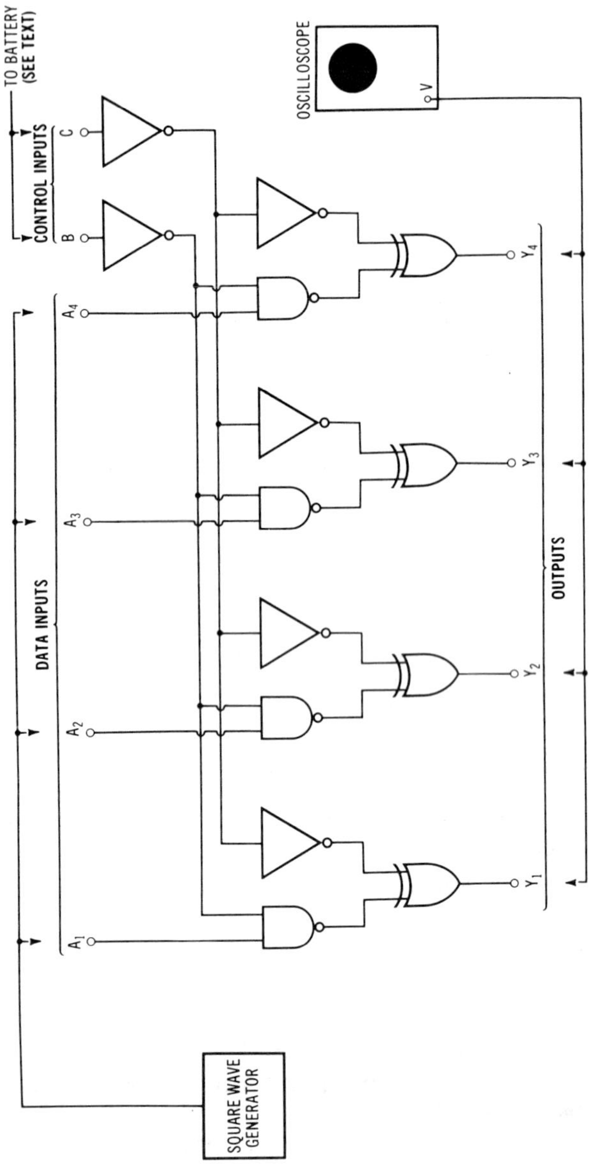

(*A*) *Test setup.*

Fig. 4-15. Operating check of a 4-bit

NOTE 4-9

True/Complement Zero/One Element Function

True/complement zero/one elements function to pass a binary number unchanged, or to complement the number in passage. Thus, 1011 can be outputted as 1011, or it can be outputted as 0100. This 1011 output is called the true mode of operation, whereas the 0100 output is called the complementary mode of operation. A true/complement function is basic in arithmetic and logic units (alu's), because addition occurs when the data is true, and subtraction occurs when the data is complementary. All arithmetical operations are performed by adders in conjunction with true/complement elements and storage devices such as shift registers.

4-12. To Test the Operation of a Parity Checker/Generator

Equipment: Square-wave generator, and battery.

Connections Required: Refer to Fig. 4-16. Connect output from generator to each of the input terminals on the device in turn; then connect generator output to pairs of the device input terminals in turn. Connect device outputs to the vertical-input channels of a dual-trace scope. Connect output enable terminal in turn to a logic-high source and to a logic-low source. Use rated V_{cc} supply voltage.

Procedure: As each device input is driven by the square-wave signal, observe scope screen as the enable terminal is switched from logic-high to logic-low. Repeat as various pairs of device input terminals are driven by the generator. (See diagram.)

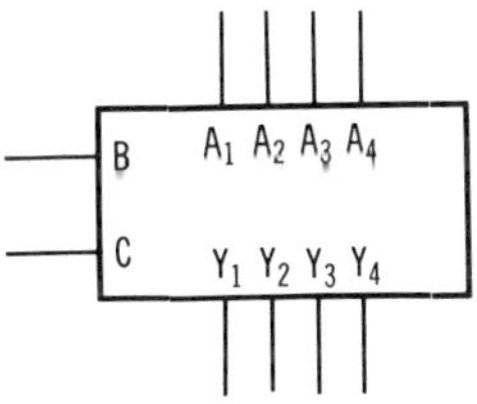

(B) Basic true/complement zero/one element symbol.

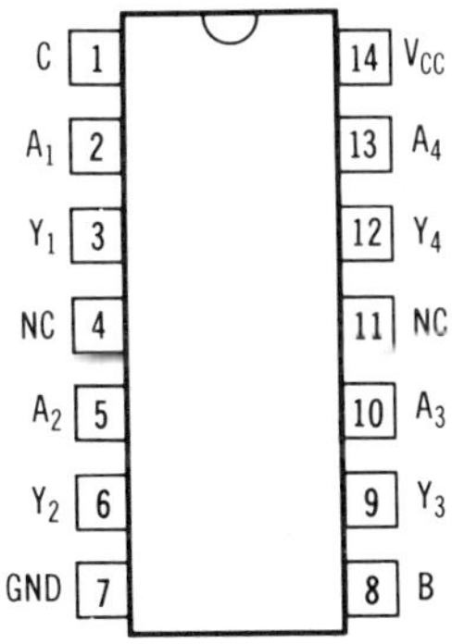

(C) Typical package pin-out.

CONTROL INPUTS		OUTPUTS			
B	C	Y_1	Y_2	Y_3	Y_4
0	0	$\bar{A}_1$	$\bar{A}_2$	$\bar{A}_3$	$\bar{A}_4$
0	1	A_1	A_2	A_3	A_4
1	0	1	1	1	1
1	1	0	0	0	0

(D) Truth table.

true/complement zero/one element.

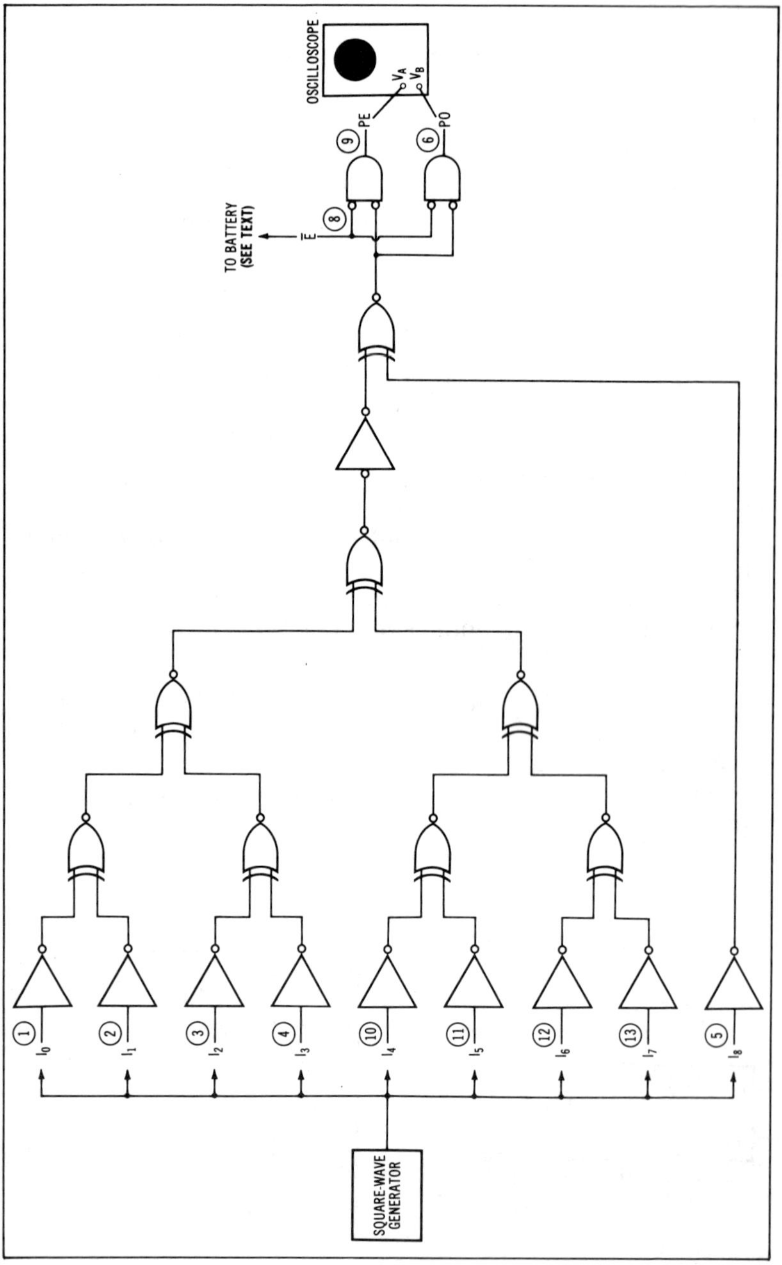

(*A*) *Test setup.*

Fig. 4-16. Test of parity

Evaluation of Results: In normal operation, an output square wave will be obtained from the PO terminal when one input terminal is driven and the enable terminal is logic-low; no output square wave is normally displayed when the enable terminal is logic-high. Similarly, an output square wave is normally obtained from the PE terminal when any two input terminals are driven and the enable terminal is logic-low; no output square wave is normally displayed when the enable terminal is logic-high.

NOTE 4–10

Even and Odd Parity

Parity checking is utilized in routine error checks of data processing operations. The exemplified device provides odd and even parity checks for digital words up to nine bits. The even parity output, PE, is normally logic-high if an even number of logic 1's are applied to the device inputs. On the other hand, the odd parity output, PO, is normally logic-high if an odd number of logic 1's is applied to the device inputs. Note that the enable input, E, normally forces both outputs to a logic-low level when a high level is applied to the enable input terminal.

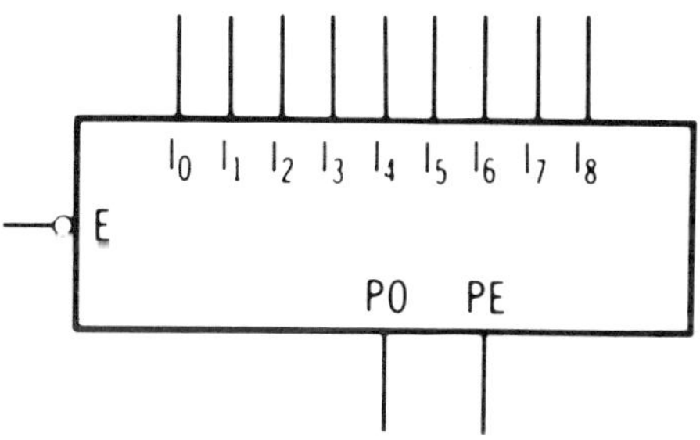

(*B*) *Basic parity checker/generator symbol.*

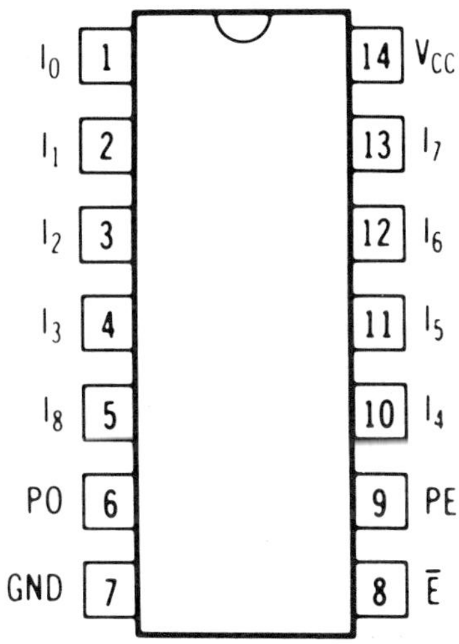

(*C*) *Typical package pin-out.*

checker/generator.

SECTION 5

Semiconductor Tests and Measurements

5–1. To Test a Diode Junction With a Quick Checker

Equipment: Quick checker arrangement as shown in the diagram of Fig. 5-1.

Connections Required: Connect output leads from quick checker to scope, as shown. Connect test leads of quick checker across the diode junction under test.

Procedure: Adjust scope gain controls for suitable size of pattern. Observe shape of displayed pattern (Fig. 5-2).

Evaluation of Results: A normal diode will display a right-angled pattern on the scope screen. There may be either one or two right angles, depending on the diode type. The first right angle is the result of forward conduction in the diode. The second right angle (if present) is the result of zener action by the diode.

NOTE 5–1

Operation of Quick Checker

The 10-to-1 stepdown transformer used in the quick checker can be quite small, because the maximum secondary current flow (with red and black test leads short-circuited) is only 11 mA, approximately. The open-circuit output voltage between the leads is approximately 12 V rms; thus, when the leads are shorted, the drop across the 1K resistor is approximately 30 V p-p. Current flow through the device under test is limited

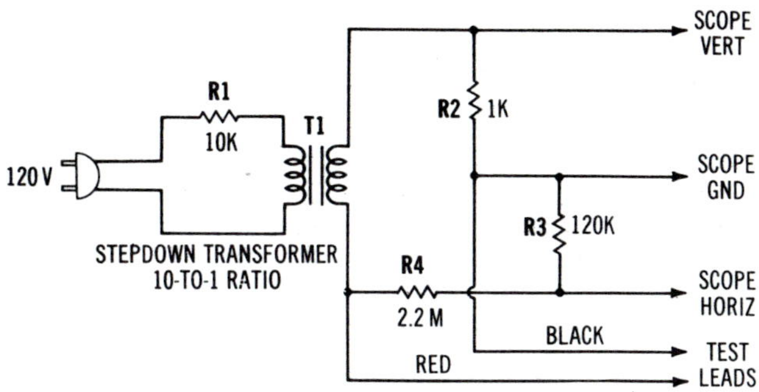

(*A*) *Circuit of quick checker.*

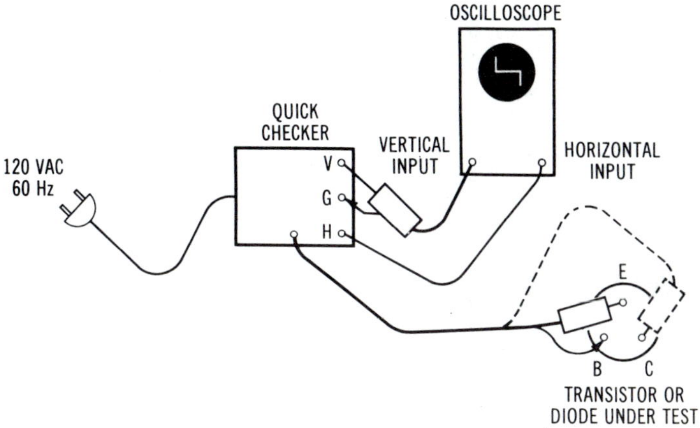

(*B*) *Application of test leads.*

Fig. 5-1. Test of diode junction action.

by R1, by R2, and by the internal resistance of the device. A normal junction has low forward resistance and high back resistance. This difference in junction resistance values produces the characteristic right-angled pattern on the scope screen. If a right angle does not appear in the pattern, the diode is defective and should be replaced. *Note that the apparent right angle is actually a section of an exponential curve. However, the diode is swept over a comparatively large voltage interval, with the result that the exponential curve is highly compressed in the pattern, and has the appearance of a right angle.*

5–2. To Estimate Resistance Values With a Quick Checker

Equipment: Quick checker.

Connections Required: Connect quick checker to scope, as shown in Fig. 5-1. Connect test leads from quick checker across resistor under test.

Procedure: Observe the slope of the diagonal line displayed on the scope screen.

Evaluation of Results: Typical pattern slopes produced by resistance values from 100 ohms to 10 kilohms are shown in Fig. 5-3. A "shorted" diode junction never has zero ohms resistance, although its value might be comparatively low. A short-circuited junction will not produce a right-angled pattern; however, if its internal resistance is appreciable, a 45° angle might be produced, for example. In any case, the slopes will be proportional to the resistance values. For precise calibration of slopes, the test leads from the quick

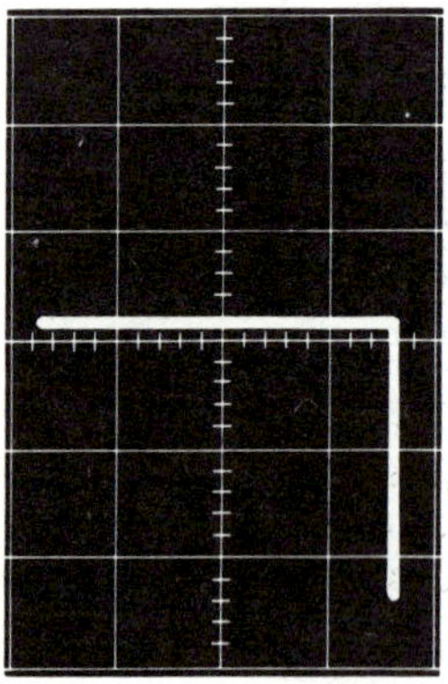

(A) Black test lead connected to diode cathode.

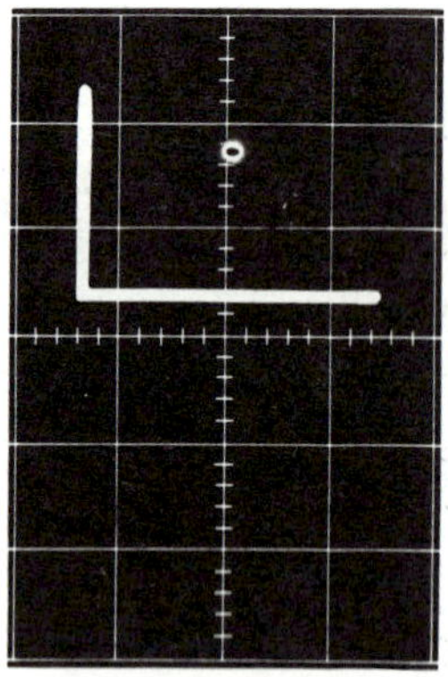

(B) Black test lead connected to diode anode.

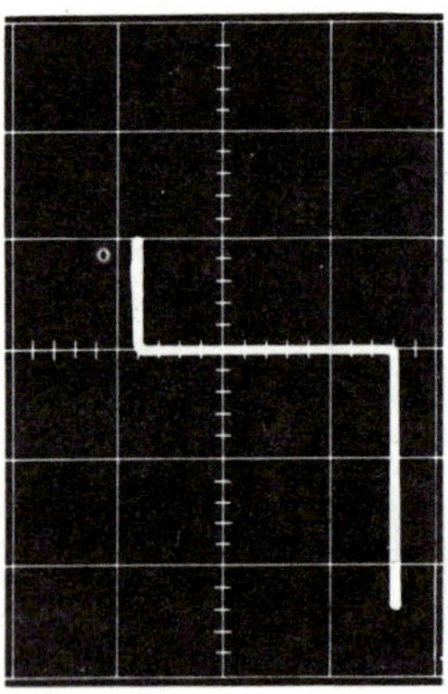

(C) Waveform of a 30-volt zener diode.

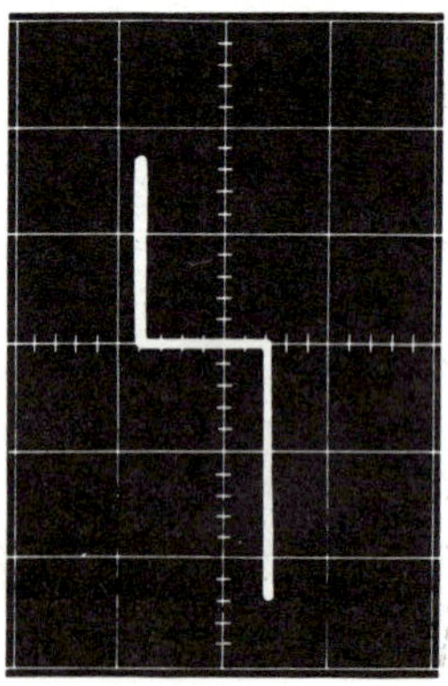

(D) Waveform of a 15-volt zener diode.

Fig. 5-2. Normal waveforms displayed

checker can be applied to precision resistors, and the scope controls adjusted for a desired pattern aspect.

5–3. To Check a Bipolar Transistor

Equipment: Quick checker.

Connections Required: With reference to Fig. 5-1, connect output leads from quick checker to scope. Connect test leads from quick checker to transistor terminals as explained next.

Procedure: (Silicon transistors). Apply the quick-checker red and black test leads at random to the transistor terminals. When a waveform is obtained as exemplified in Fig. 5-4, the leads will be connected across the base-emitter junction of the device. Next, connect the red test lead to the collector terminal. Then connect the black test lead alternately to the two remaining terminals of the transistor. A greater vertical deflection will be obtained when the black test lead is connected to the base of the transistor.

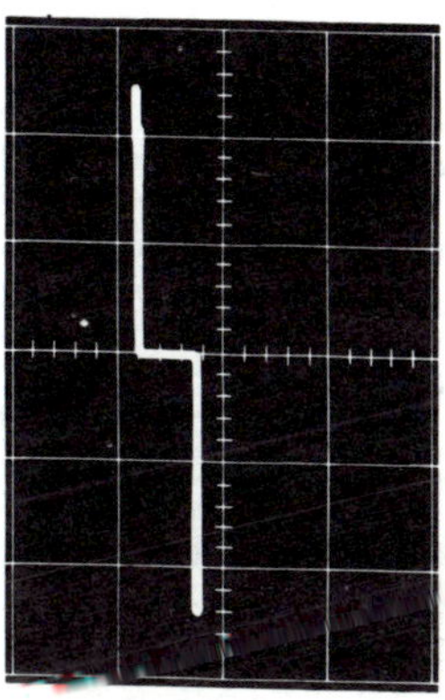

(*E*) *Waveform of a 7-volt zener diode.*

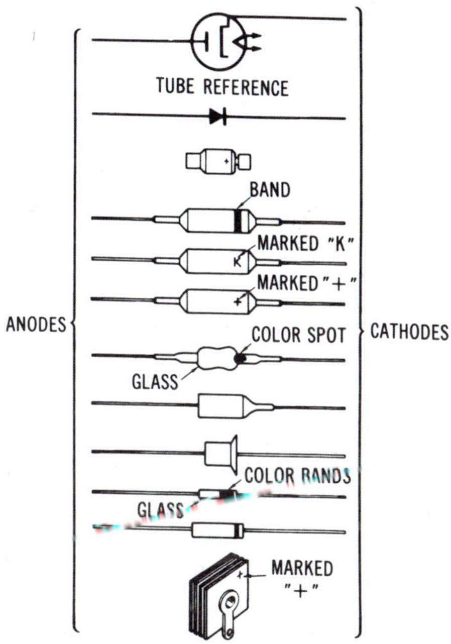

(*F*) *Diode polarity identifications.*

when the test leads are connected to a diode.

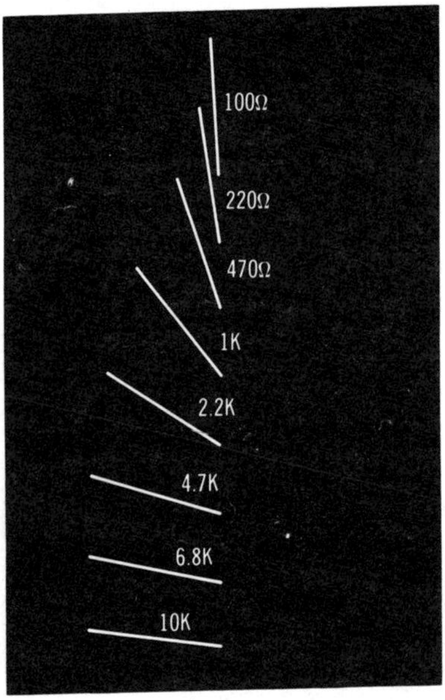

Fig. 5-3. Slopes of screen traces for corresponding values of resistances.

Evaluation of Results: The transistor is workable if the foregoing test results are obtained; the base terminals will also be identified by the test results. In the final step, vertical deflection will be downward at the right and upward at the left for an npn transistor. Deflection is opposite for a pnp type.

NOTE 5–2

Silicon and Germanium Transistor Characteristics

Most transistors are silicon types, and have zener-type conduction through their base-emitter junction; this results in a pattern that has both upward and downward deflection intervals. However, germanium transistors do not have zener conduction, and the foregoing first procedural test cannot be carried out. Hence, the following procedure is used instead: Connect the red and black test leads at random to the transistor terminals. When an open circuit (horizontal line) is displayed on the scope screen, the test leads are connected to the collector and emitter (a small conduction may be observed at the voltage crossover point. Next, connect the black test lead to the base terminal of the transistor. Observe the scope screen; a pnp transistor produces downward deflection at the right, and an npn transistor produces upward deflection at the left of the pattern. (A germanium transistor does not produce both upward and downward deflection intervals.)

5–4. To Check a Bipolar Transistor With a Semiconductor Curve Tracer

Equipment: Curve tracer as depicted in Fig. 5-5.

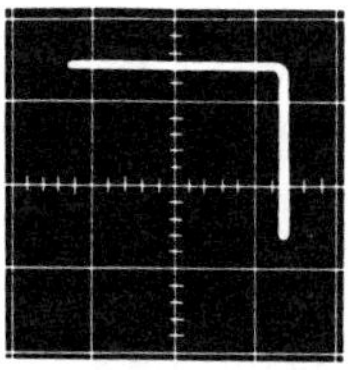

(*A*) *Pnp base-emitter waveform.*

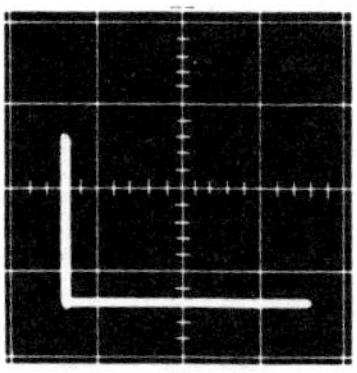

(*B*) *Npn base-emitter waveform.*

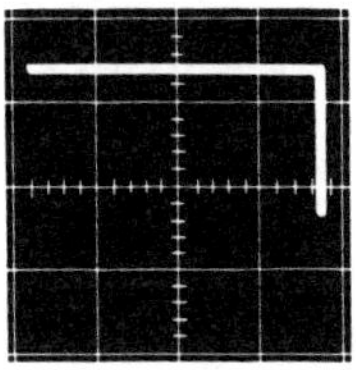

(*C*) *Ppn collector-base waveform with red lead connected to collector.*

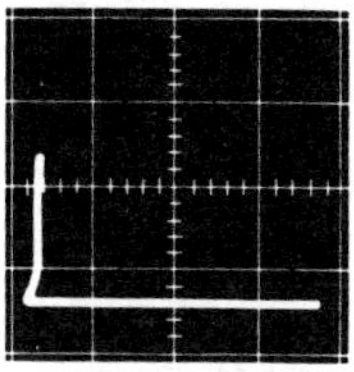

(*D*) *Npn collector-base waveform with red lead connected to collector.*

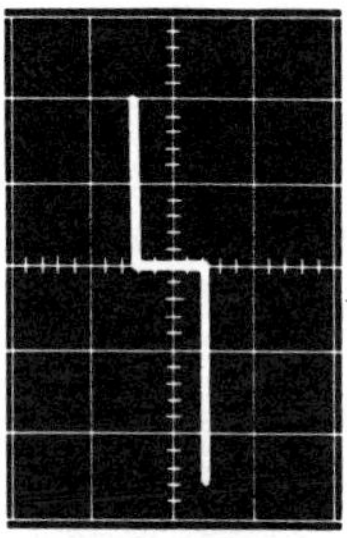

(*E*) *Black lead connected to base* (*pnp*).

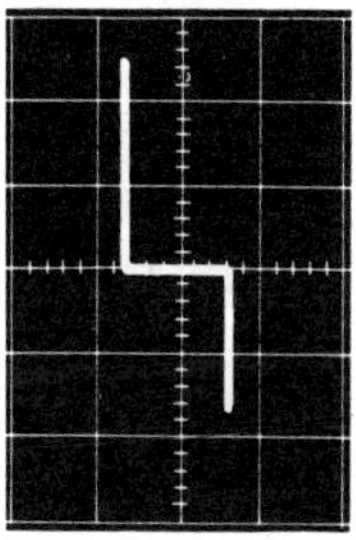

(*F*) *Black lead connected to base* (*npn*).

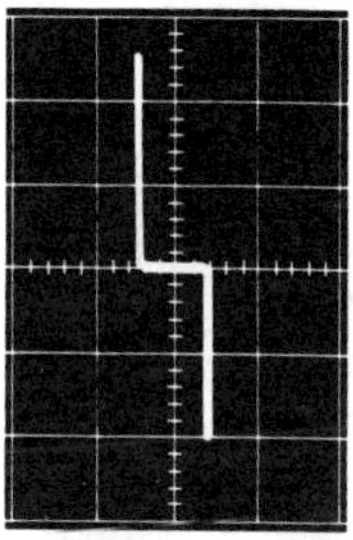

(*G*) *Black lead connected to emitter* (*pnp*).

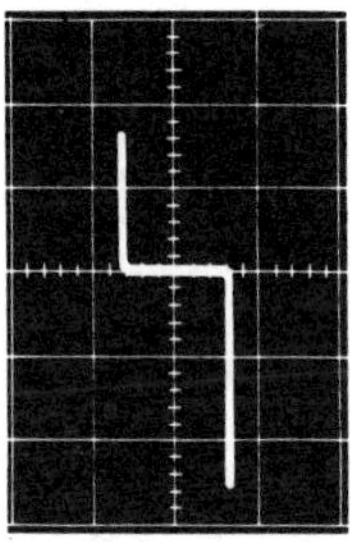

(*H*) *Black lead connected to emitter* (*npn*).

Fig. 5-4. Waveforms normally produced by transistors.

Connections Required: Connect output leads from tracer to the vertical and horizontal input terminals of the scope. Plug transistor under test into the socket provided on the tracer.

Procedure: Set the bias step-voltage control as required for the type of transistor under test. Adjust scope gain controls for suitable size of pattern.

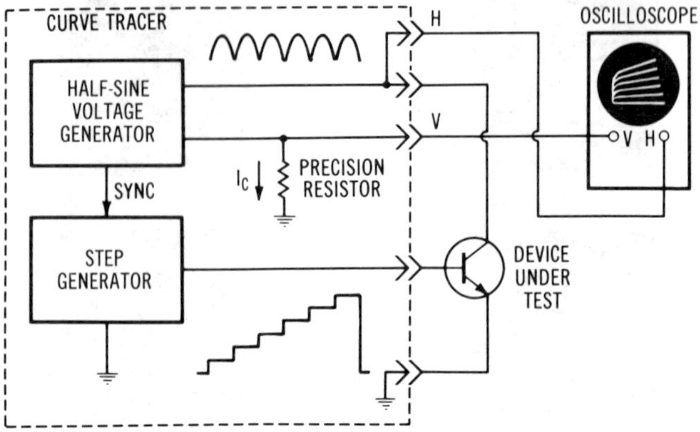

(*A*) *Test setup.*

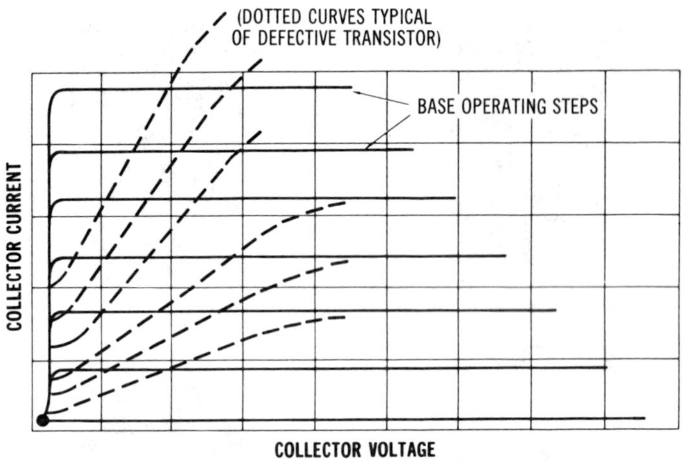

(*B*) *Typical six-step pattern for a silicon transistor.*

Fig. 5-5. Display of collector-family characteristics.

Evaluation of Results: A family of collector characteristics will be displayed on the scope screen if the transistor is in workable condition. A silicon transistor with very small reverse collector current (very high back resistance) will produce a family of traces that are nearly horizontal. Collector leakage causes the curves to slope uphill on the screen. Low gain causes the family of curves to be compressed at low collector voltage, and to be expanded at high collector voltage. A germanium transistor normally has lower back resistance than a silicon transistor. In turn, characteristics for germanium transistors tend to slope uphill on the screen to some extent. In case of doubt, compare the test pattern with that produced by a similar known good transistor (or consult the manufacturer's data sheet).

NOTE 5-3

Nonlinear Resistance Indication

Observe that the basic distinction between the quick checker described in Fig. 5-1 and the curve tracer exemplified in Fig. 5-4 is that the latter provides a series of step voltages to obtain a multitrace display. In turn, the slope of a curve in Fig. 5-5 indicates a resistance value, just as the slopes of the traces in Fig. 5-3 represent resistance values. The slope of a curve indicates the resistance from the collector terminal to the emitter terminal of the transistor under test. Note that the traces are seldom precisely straight, but tend to be curved more or less. This curvature corresponds to nonlinear resistance (resistance wherein the current flow is not directly proportional to the applied voltage). Zero or infinite resistance values (shorted or open transistors) produce approximately vertical or horizontal line patterns, as illustrated in Fig. 5-6. (See also Note 5-4.)

5-5. To Measure the Beta Value (Current Gain) of a Transistor

Equipment: Same as in Fig. 5-4.

Connections Required: Same as in Fig. 5-4.

Procedure: Calibrate the scope with the controls provided on the curve tracer, so that a suitable range such as 10 volts is indicated on the horizontal axis, and a suitable range such as 30 milliamperes is indicated on the vertical axis. Position the pattern as shown in Fig. 5-7.

Evaluation of Results: As illustrated in the diagram, the vertical spacing of successive traces corresponds to a collector-current interval termed ΔI_C. The base-current steps provided by the curve tracer are set to a base-current interval termed ΔI_B. In turn, the current gain (beta) at a chosen area in the pattern is given by $\Delta I_C/\Delta I_B$. In the illustrated example, ΔI_C is measured at a region in the pattern where its value is 9 mA, since the value of ΔI_B has been set to 0.05 mA, the corresponding value of beta is 180.

NOTE 5-4

Display of NPN and PNP Characteristics

Since npn- and pnp-type transistors conduct in opposite directions, a pnp family of characteristics will be displayed "upside down" with respect to an npn family of characteristics, as depicted in Fig. 5-8. Thus, an npn display is zeroed at the lower left-hand corner of the graticule,

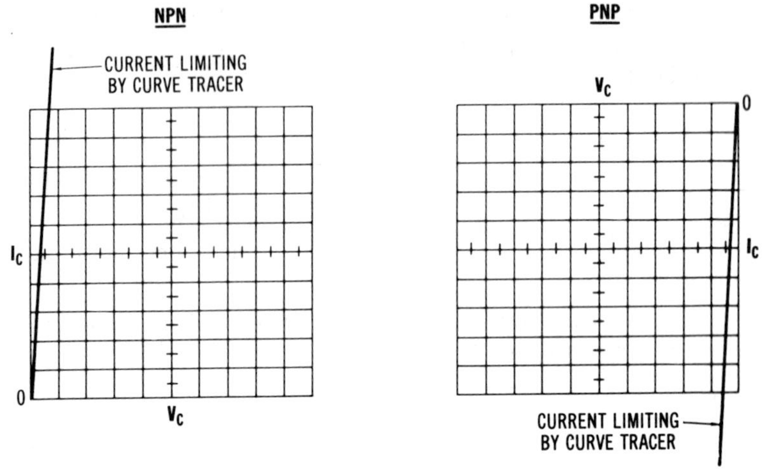

(*A*) *Shorted transistor.*

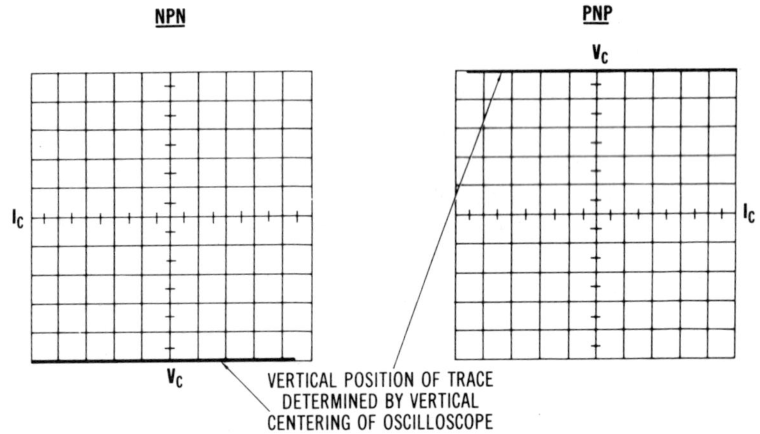

(*B*) *Open transistor.*

Fig. 5-6. Transistor curves.

whereas a pnp display is zeroed at the upper right-hand corner of the graticule. Note that the scope should have dc-coupled amplifiers; ac-coupled amplifiers will cause trace shift and distort the display. Also, it is desirable that the horizontal amplifier have very high input resistance, because this input resistance shows up in the pattern as apparent transistor leakage resistance. To observe this apparent leakage, switch the transistor in and out, watching for any change (movement) in the base line.

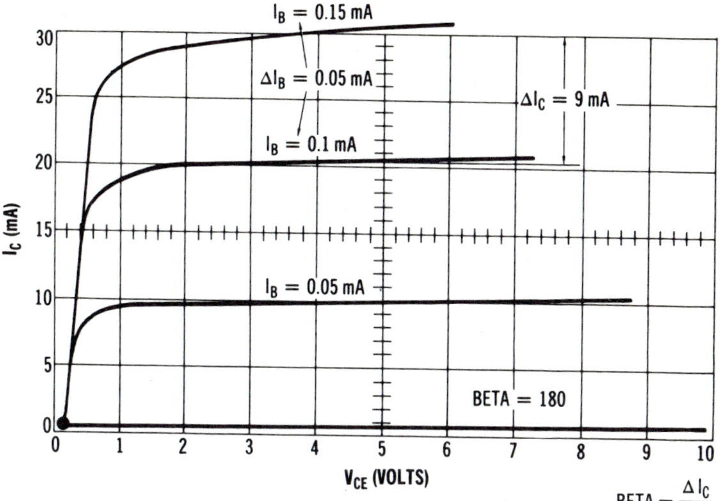

Fig. 5-7. Determination of transistor beta value.

5–6. To Measure the Output Impedance (Resistance) of a Transistor

Equipment: Same as in Fig. 5-4.

Connections Required: Same as in Fig. 5-4.

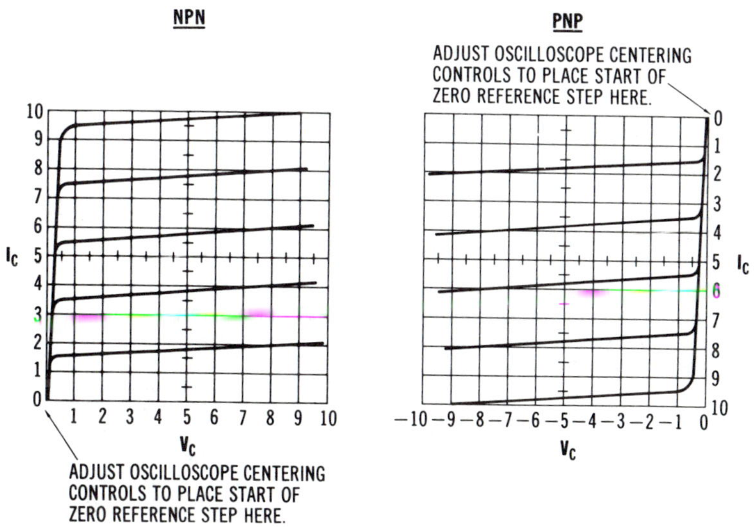

Fig. 5-8. Pnp display is "upside down" with respect to npn display.

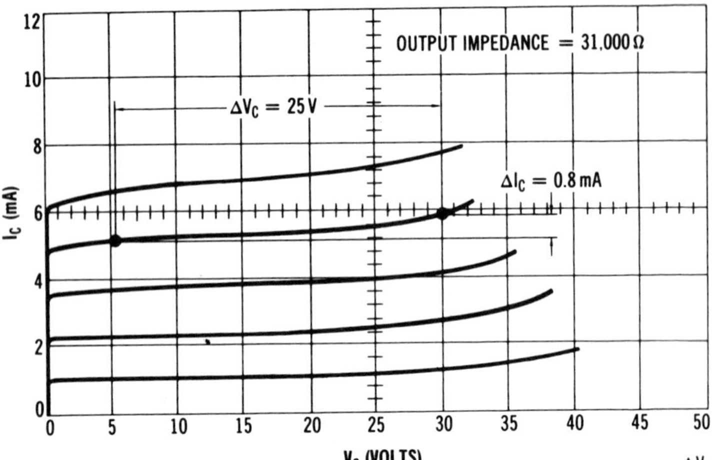

$$\text{OUTPUT IMPEDANCE} = \frac{\Delta V_C}{\Delta I_C}$$

Fig. 5-9. Determination of transistor output impedance.

Procedure: Calibrate the scope for suitable voltage and current ranges such as 50 volts on the horizontal axis and 12 milliamperes on the vertical axis. Position the pattern as shown in Fig. 5-9.

Evaluation of Results: As indicated in the diagram, the slope of a trace corresponds to a collector-current interval termed ΔI_C. This collector-current interval has an associated collector-voltage interval termed ΔV_C. In turn, the output impedance denoted by these current and voltage values is given by $\Delta V_C/\Delta I_C$. In the illustrated example, this ratio is equal to 25/0.0008, or 31,000 ohms. *As a general rule, power-type transistors have lower output impedance than small-signal transistors.*

NOTE 5–5

Looping and Thermal Runaway

Test procedures should always be conservative, so that the device under test is not overdriven. Conduction of current through a transistor generates heat. If the collector voltage and/or base current results in excessive heating, the traces that are normally displayed change into loops. The size of the loop increases as the amount of overdrive is increased. This condition is incipient "thermal runaway" which will destroy a transistor unless the drive is decreased. Overdrive causes looping to appear in the pattern because the transistor heats up rapidly as the peak of the drive voltage or current is approached, and then cools to some extent as the peak is passed. Inasmuch as heating occurs before cooling, there is a time lag involved that shows up in the pattern as a looping.

5-7. To Check the Condition of a Transistor In-Circuit

Equipment: Transistor curve tracer.

Connections Required: Remove power from equipment under test; then connect the curve-tracer test leads to the transistor on the printed-circuit board.

Procedure: Since circuit impedances may shunt considerable base-drive voltage around the transistor, it may be necessary to apply more base-drive step voltage than in out-of-circuit tests. Occasionally, all of the available drive signal may be shunted away by a very low in-circuit base-emitter resistance, and no curve display can be obtained on the scope screen.

Evaluation of Results: In-circuit patterns generally appear highly distorted, as compared with out-of-circuit collector characteristic patterns (see Fig. 5-10). This distortion is due to in-circuit impedance. However, it is possible to interpret various displays. For example, five loops and a base line appear in the npn pattern. This is interpreted to mean that every base-voltage step produces additional collector current, and that the transistor is probably workable. The pnp pattern is typical of an oscillator circuit; it seems to indicate serious transistor leakage, although the transistor is normal. In summary, in-circuit pattern evaluation is best accomplished on the basis of comparative tests, or on the basis of experience with specific circuitry.

NOTE 5-6

Impedance Loops Versus Thermal Loops

The loops shown in Fig. 5-10 are typical impedance loops, caused by RC or RL circuitry. It is easily possible to distinguish between such im-

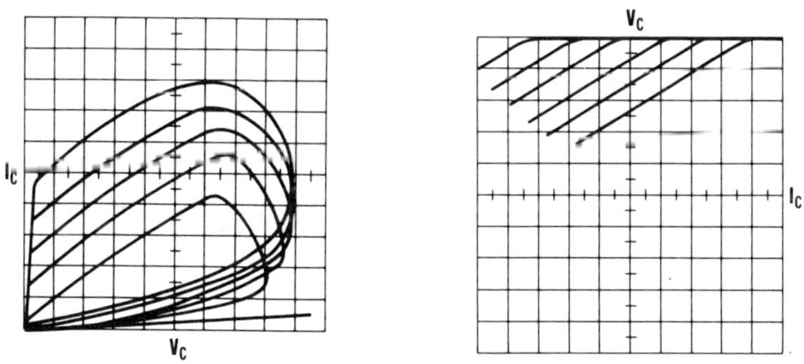

(A) *For an npn transistor.* (B) *For a pnp transistor.*

Fig. 5-10. Typical in-circuit curve-tracer patterns.

pedance loops and the thermal loops described in Note 5-5. Thus, the loop size decreases and disappears as the drive is decreased, in the case of a thermal loop. On the other hand, an impedance loop does not disappear as the drive is decreased.

5–8. To Select a Pair of Matched Transistors

Equipment: Transistor curve tracer.

Connections Required: Insert the various transistors from an assortment successively into the test sockets on the curve tracer.

Procedure: Observe the screen patterns produced by the chosen transistors.

Evaluation of Results: A pair of matched transistors will produce patterns that are virtually identical. In complementary-symmetry amplifiers, a pnp transistor is to be matched by an npn transistor. Therefore, the test patterns will have opposite polarity, as depicted in Fig. 5-11. However, the pnp pattern matches the npn pattern if it is a mirror image.

NOTE 5–7

"Vertical Roll" in Loop Pattern

If excessive heat is generated in a transistor while testing with a curve tracer, the condition becomes apparent in the scope pattern (Fig. 5-12). Looping appears in the traces, because the collector current does not increase and decrease at the same rate as the sweep voltage increases and decreases. As the sweep voltage starts from zero, the transistor is comparatively cool. The transistor temperature increases to a maximum at the peak of the sweep voltage, and increasing temperature causes addi-

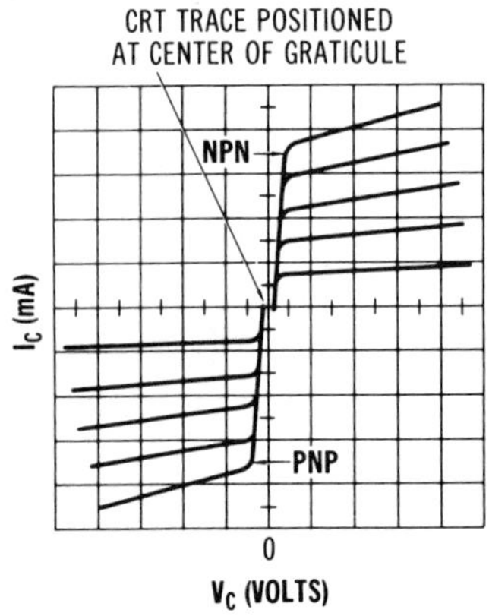

Fig. 5-11. Mirror-image patterns indicating that the npn transistor matches the pnp transistor.

tional collector-current flow and additional heat. Then, as the sweep voltage decreases, a time lag is involved in cooling of the transistor. The top portion of a loop shows the increasing sweep current, and the bottom portion of the loop shows the decreasing sweep current. Note also that with some transistors, the collector current will droop at the high end of the curve—an increase in temperature is causing a decrease in collector current. If thermal runaway starts, "vertical roll" appears; the entire family of curves moves in the direction of higher collector current, and can soon result in transistor burn-out.

5–9. To Check Nonlinearity in Transistor Current Gain

Equipment: Transistor curve tracer.

Connections Required: Same as in 5–4.

Procedure: Observe the screen pattern produced by the transistor under test.

Evaluation of Results: With reference to Fig. 5-13, plot an imaginary line along the ends of the curves; this is the "test load line." Next, plot an "operating load line" in parallel with the test load line, but intersecting the zero I_C line at the desired operating V_{CE} for the transistor. Then, measure and compare the changes in collector current (ΔI_C) between the curves on operating load line. If these changes are the same, the transistor is linear in the test region. On the other hand, if these changes are not the same, the transistor is nonlinear in the test region. In Fig. 5-14, there is some nonlinearity displayed. In other words, because Δ9mA is unequal to Δ11mA, amplitude distortion will occur in the test region. For example, an input signal of ±0.05 mA will produce an output signal of +9 mA and −11 mA.

NOTE 5–8

Variation in Nonlinearity

Transistor gain is not entirely constant, and may be quite nonlinear. The amount of nonlinearity is dependent upon the region of measurement in the pattern. Nonlinearity may be desirable or undesirable, depending upon the application of the transistor. Nonlinearity should be measured along a load line, as has been explained; this method more nearly duplicates circuit operating conditions than if the nonlinearity is measured at a specific value of V_{CE}. A load resistance causes circuit operation along a load line, because a change in collector current produces a change in voltage drop across the load, and in turn produces a change in collector voltage.

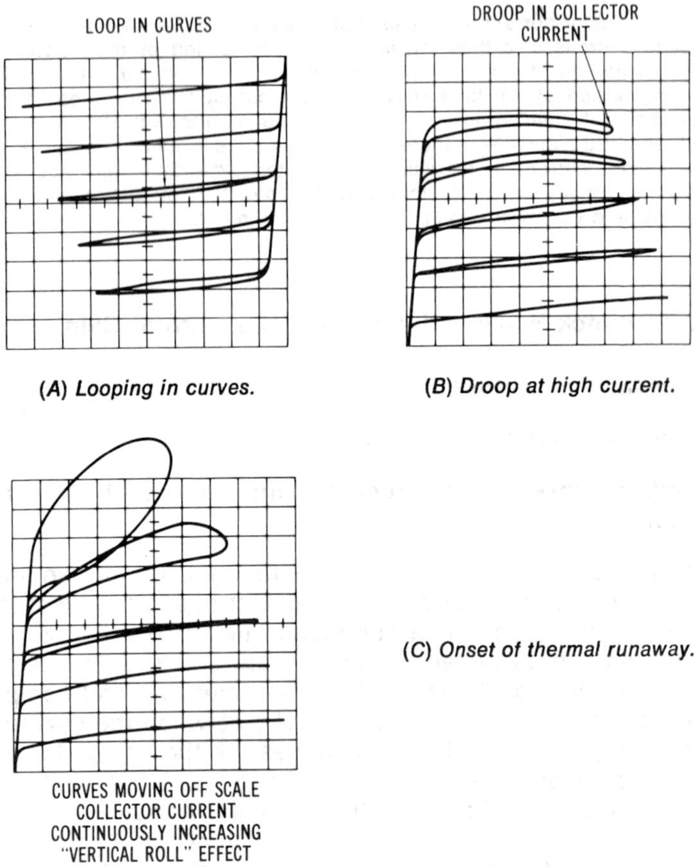

Fig. 5-12. Evidence of transistor heating in scope patterns.

5–10. Transistor Breakdown Voltage Measurement

Equipment: Transistor curve tracer.

Connections Required: Same as in 5–4.

Procedure: Display the collector family on the scope screen, with reduced horizontal width. Then, increase the sweep-voltage control on the curve tracer until an upturn in collector current is observed at the tail of the curves. (See Fig. 5-15.)

Evaluation of Results: The upturn in collector current at the onset of breakdown is very sharp for most transistors, although it is somewhat gradual for a few types. Read the collector

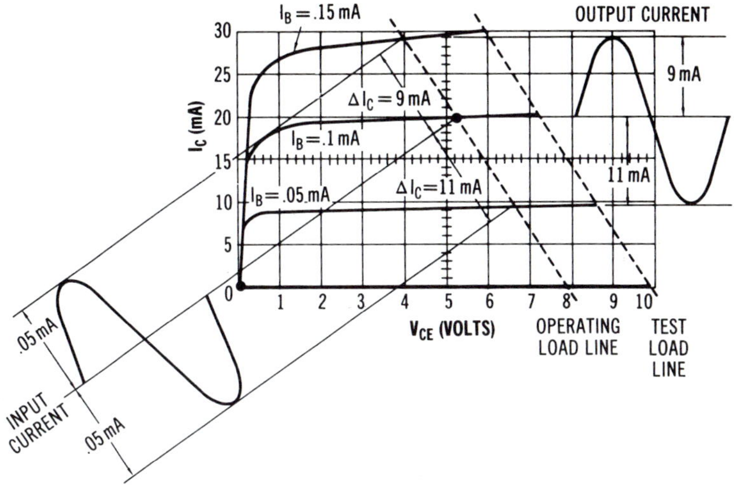

Fig. 5-13. Measurement of nonlinearity in transistor characteristics.

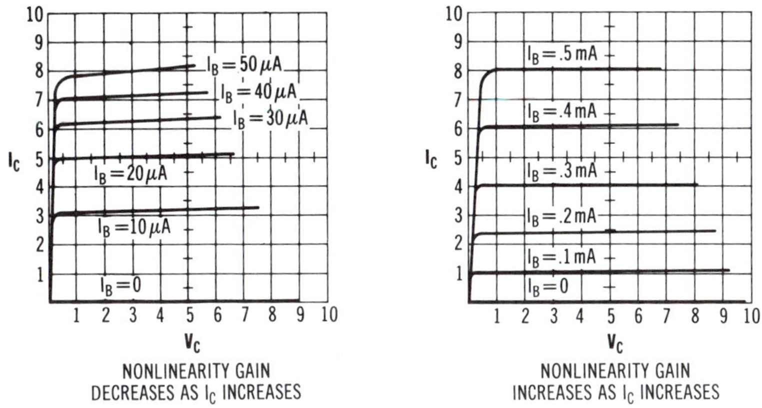

Fig. 5-14. Two types of nonlinear collector family characteristics.

voltage value at which the curve upturn starts along the calibrated horizontal axis.

NOTE 5–9

Breakdown Testing Technique

The measured breakdown voltage will normally be greater than the breakdown voltage rating in the transistor data sheet. Otherwise, the transistor is defective and should be replaced. If the breakdown voltage rating is not available, a practical rule of thumb is that the breakdown voltage should be double the value of the collector supply voltage of the circuit

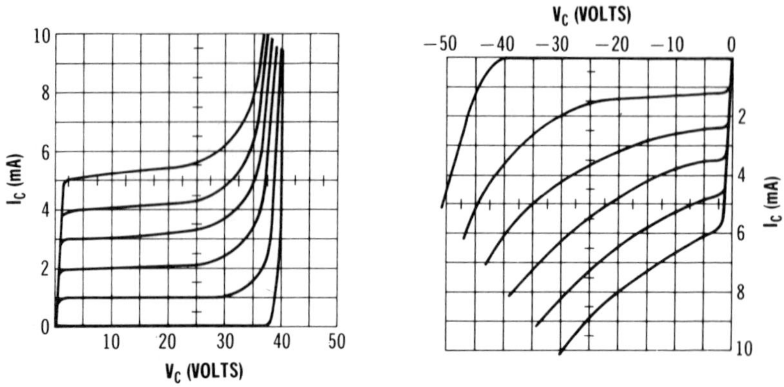

(*A*) *Abrupt breakdown.* (*B*) *Gradual breakdown.*

Fig. 5-15. Patterns observed during breakdown voltage measurements.

in which the transistor is to be used. *Breakdown voltage measurements should be made carefully, so that current upturn is not substantial, with the associated danger of thermal runaway and transistor burnout. For the same reason, breakdown voltage tests should not be prolonged.*

5–11. Transistor Saturation Voltage and Resistance Measurement

Equipment: Transistor curve tracer.

Connections Required: Same as in 5–4.

Procedure: Display the collector family and note that the knee of each curve (Fig. 5-16) occurs at approximately the same collector voltage, regardless of base current.

Evaluation of Results: The value of the saturation voltage $V_{CE(sat)}$ is defined as the collector voltage at the knee of a curve. The saturation resistance $r_{CE(sat)}$ is equal to V_C/I_C; it is equal to the collector voltage divided by the collector current for a given value of base current in the collector saturation region.

NOTE 5–10

Saturation Voltage and Resistance Ratings

For measurement of saturation voltage in comparison to data-sheet specifications, the base current and collector current should be noted at the point of measurement. The manufacturer's rating is the maximum value at which the knee should occur. In other words, if the rating is on or above the knee, the transistor is acceptable. Note that to measure saturation voltage on the curve tracer, only the saturation region of the characteristics need be displayed; this is the low collector-voltage re-

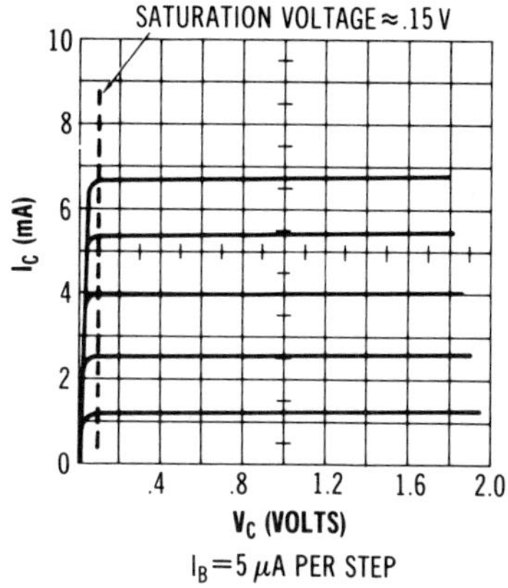

Fig. 5-16. Measurement of saturation voltage.

gion up to and including the knee of each curve. Thus, the display can be expanded, using a low-voltage horizontal calibration value, such as 0.2 volt per division. This facilitates accurate measurement of low collector voltage values. Data sheets usually specify saturation-resistance values as maximum acceptable limits.

5–12. To Display a Drain Family of FET Characteristic Curves

Equipment: Transistor curve tracer.

Connections Required: Same as in 5–4.

Procedure: Keep one hand grounded while handling the FET with the other hand, to avoid device damage from static electricity. Set tracer controls for FET test. Ground the shield of the FET (if present). In the case of a dual-gate device, ground the gate which is not being driven in the test. Be careful not to overdrive and damage the FET.

Evaluation of Results: Compare the drain family displayed in the screen pattern with the curves specified in the device data sheet. A normal device will have rated tolerances. If the FET is out of tolerance, it is defective and should be replaced (Fig. 5-17).

NOTE 5–11

Depletion and Enhancement Types of FETs

FET characteristic curves show the relation of drain current versus drain voltage at various gate voltages. FET breakdown voltage may be checked and measured in the same manner as for bipolar transistors. When checking FETs, a curve tracer is set to apply constant-voltage steps to the gate, instead of constant-current steps as in the case of

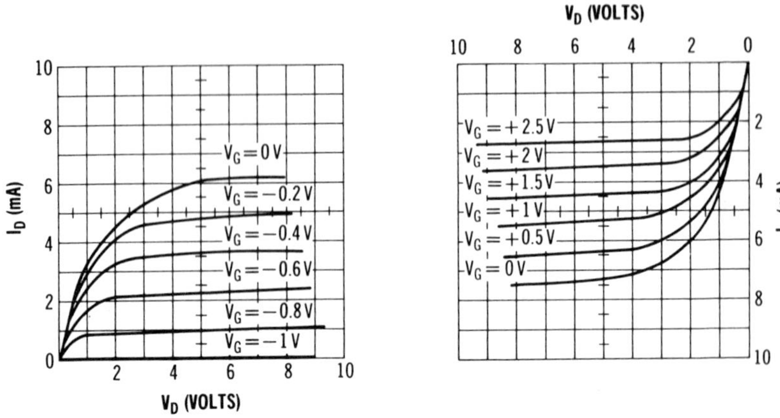

(*A*) *N-channel FET.*

(*B*) *P-channel FET.*

Fig. 5-17. Typical FET drain family characteristics.

base drive to a bipolar transistor. Most FETs are of the depletion type, and the zero-volt step corresponds to the highest drain current in the pattern. A few FETs are designed as enhancement devices, which normally have zero drain current at zero gate voltage. To check an enhancement FET, a suitable forward bias voltage must be connected in series with the gate-source lead. In turn, the enhancement device can be tested in the same manner as a depletion device.

5–13. To Check the Drain Family Characteristics for a Dual-Gate FET

Equipment: Transistor curve tracer.

Connections Required: Same as in 5–4, except that the unused gate is connected to the source. In the second part of the test, the unused gate is returned to the source through an adjustable bias voltage.

Procedure: Same as in 5–12, except that a separate test is made for each gate. Follow up with a comparison test in which the drain family characteristics are displayed with the undriven gate returned to the source through low, medium, and high values of bias voltage.

Evaluation of Results: Consult the data sheet for the device, and observe whether the FET is within rated tolerances in each part of the test procedure. Note that dual-gate FETs are used, for example, in agc circuitry and in heterodyne mixer circuitry; both gates have control of the drain current flow.

5–14. To Measure the Transconductance of an FET

Equipment: Transistor curve tracer.

Connections Required: Same as in 5–4.

Procedure: Observe the change in drain current (I_D) that results from a corresponding change in gate voltage (V_G) within the region of the characteristics chosen for measurement (Fig. 5-18).

Evaluation of Results: Transconductance is measured in mhos, and is equal to a given change in drain current divided by the associated change in gate voltage:

$$g_m = \frac{\Delta I_D}{\Delta V_G}$$

For example, in Fig. 5-18 a change of 1.1 mA in drain current is caused by a change of 0.5 volt in gate voltage. Accordingly, the transconductance of the FET in the region of measurement is 2200 micromhos (μmhos).

NOTE 5–12

Voltage- and Current-Operated Devices

The gain of an FET is measured in terms of transconductance because the gate draws negligible current; an FET is said to be a voltage-operated device. On the other hand, the gain of a bipolar transistor is measured in terms of a current ratio (beta) because the base draws current; a bipolar transistor is said to be a current-operated device. In turn,

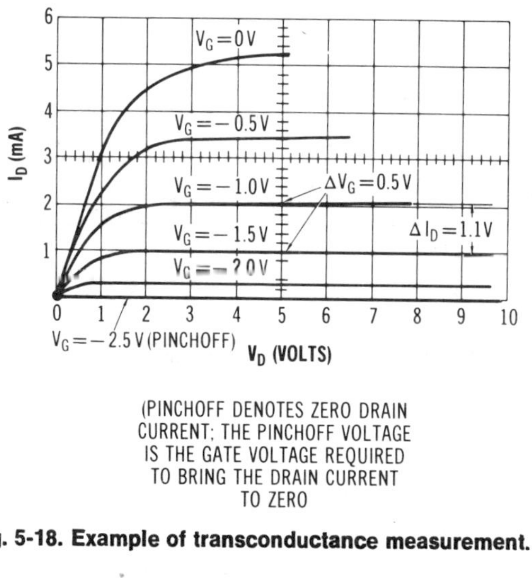

Fig. 5-18. Example of transconductance measurement.

the gain of an FET in micromhos cannot be directly compared with the gain of a bipolar transistor in beta units. However, it may be noted that the gain of a bipolar transistor can be measured in terms of transconductance, if desired. This is done by measuring the change in collector current that results from an associated change in base voltage. In the majority of practical applications, it is preferable to consider the gain of a bipolar transistor in beta units.

5–15. To Determine the Pinchoff Voltage of an FET

Equipment: Transistor curve tracer.

Connections Required: Same as in 5–4.

Procedure: Start display of the pattern with a small step voltage which provides all of the available curves (such as six curves). Then, increase the step-voltage setting until the last curve disappears (merges with the next-to-the-last curve along the zero current axis). In Fig. 5-19, this increase in step voltage causes two of the latter curves to disappear.

Evaluation of Results: The pinchoff voltage is the gate voltage that corresponds to the zero-current level in the pattern. For example, in Fig. 5-19, pinchoff occurs in the range between

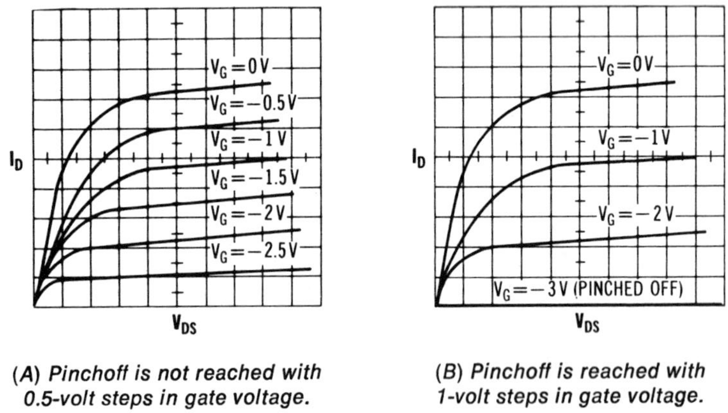

(*A*) *Pinchoff is not reached with 0.5-volt steps in gate voltage.*

(*B*) *Pinchoff is reached with 1-volt steps in gate voltage.*

Fig. 5-19. Determination of pinchoff voltage.

−2.5 and −3 volts. If a more precise measurement is desired, connect an external bias source to the gate of the FET, and adjust the bias voltage exactly to pinchoff, as indicated on the scope screen.

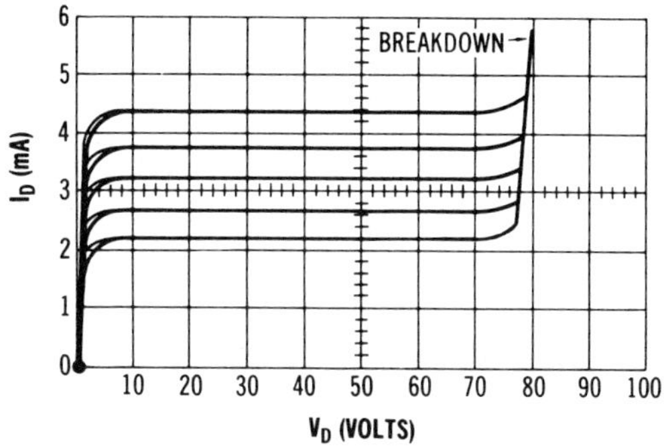

Fig. 5-20. Example of FET breakdown voltage measurement.

5–16. To Measure the Breakdown Voltage of an FET

Equipment: Transistor curve tracer.

Connections Required: Same as in 5–4.

Procedure: Display the drain family of characteristics. Increase the curve-tracer sweep voltage until the curves show the onset of breakdown (see Fig. 5-20).

Evaluation of Results: The breakdown voltage is reached at the point where the curves begin to show an abrupt upshoot in drain current. Keep the test time short, to avoid possible damage to the FET.

5–17. To Check a Silicon Controlled Rectifier for Forward Blocking Voltage

Equipment: Transistor curve tracer.

Connections Required: As depicted in Fig. 5-21, connect the SCR anode terminal to the collector jack of the curve tracer; connect the SCR gate terminal to the base jack; connect the SCR cathode terminal to the emitter jack.

Procedure: Set the step-selector control to zero volt (this is the I_{CES} position for a typical tracer); set the polarity control to npn. Then increase the sweep-voltage setting until the SCR "fires"; in other words, the anode current abruptly increases and the anode voltage drops to nearly zero.

Evaluation of Results: Read the highest anode voltage value in the display on the calibrated screen. This highest value is

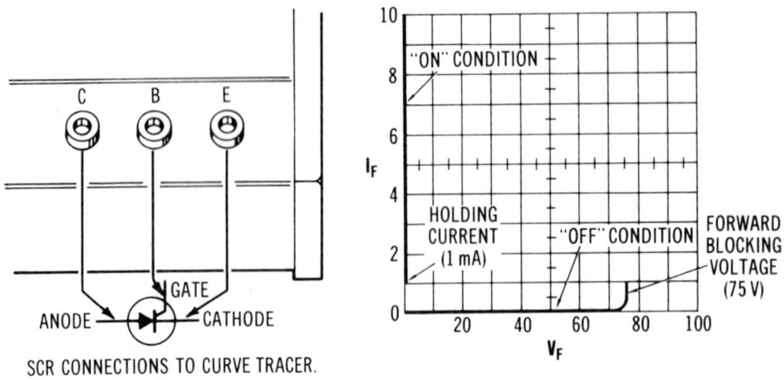

Fig. 5-21. Measurement of forward blocking voltage for an SCR; holding current is also displayed.

defined as the maximum forward blocking voltage. Any anode current at anode voltage below the firing point is forward leakage current and it can be read directly from the display.

NOTE 5–13

SCR Function

An SCR or thyristor is a four-layer pnpn device with three terminals: cathode, anode, and gate. It operates similarly to a diode in its "on" state. However, an SCR also has an "off" state in which it will not conduct in either direction. If its forward blocking voltage is exceeded, the SCR normally "turns on." It will then stay on until the anode-cathode current drops below a certain value called the holding current. The forward blocking voltage is the maximum anode-cathode voltage in the forward direction that the SCR can withstand before conduction, with the condition of zero gate current.

5–18. To Measure the Reverse Blocking Voltage for an SCR

Equipment: Transistor curve tracer.

Connections Required: Same as in Fig. 5-21.

Procedure: Same as in 5–17 except that the polarity control is set to its pnp position.

Evaluation of Results: Read the highest voltage value on the screen at which voltage breakdown occurs with an abrupt increase in anode current. Any anode current that flows at voltages below the breakdown point is reverse leakage current and can be read directly from the display.

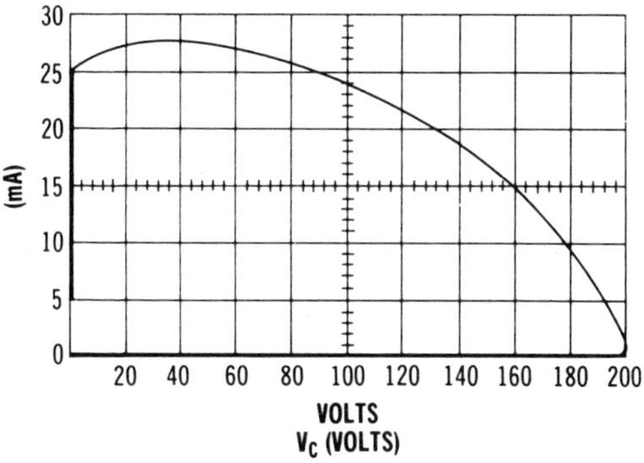

Fig. 5-22. Example of forward conduction at 200 volts, followed by holding-current indication of 5 mA.

5–19. To Measure the Holding Current for an SCR

Equipment: Transistor curve tracer.

Connections Required: Same as in Fig. 5-21.

Procedure: Same as in 5–17.

Evaluation of Results: Observe the lowest current value displayed for the "on" condition of the SCR. This is defined as the holding current. Note that the holding current can also be measured with the step-selector control set to one of its "current per step" positions, so that less sweep voltage is required to drive the SCR into its "on" condition (Fig. 5-22).

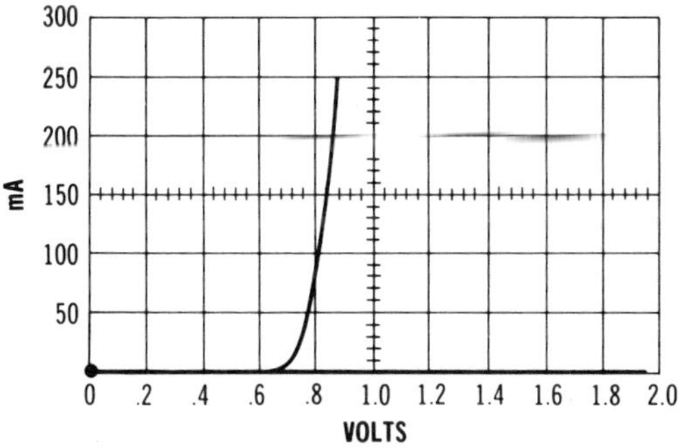

Fig. 5-23. Example of forward voltage drop display for an SCR.

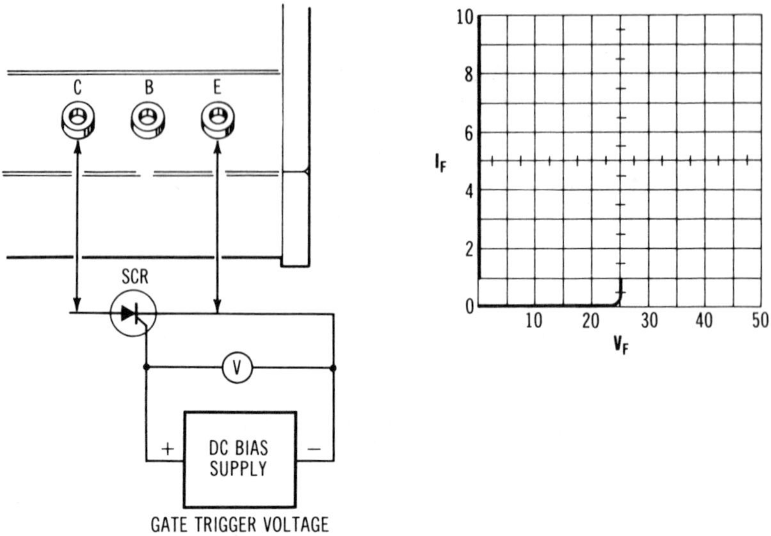

Fig. 5-24. Check of gate trigger voltage for an SCR.

5–20. To Measure the Forward Voltage Drop of an SCR

Equipment: Transistor curve tracer.

Connections Required: Same as in Fig. 5-21.

Procedure: Increase the horizontal sensitivity of the oscilloscope and display a low voltage portion of the forward voltage (Fig. 5-23). Then increase the current-per-step control setting so that the sweep voltage may be reduced. The vertical sensitivity of the scope may be reduced to 10 mA per division for a greater voltage/current range.

Evaluation of Results: Read the forward voltage value directly from the display.

5–21. To Measure the Gate Trigger Voltage for an SCR

Equipment: Transistor curve tracer, dc bias supply with meter.

Connections Required: Connect equipment as shown in Fig. 5-24.

Procedure: Same as in 5–17, but set the sweep voltage to a specified forward anode-cathode voltage, and increase the dc bias value until the SCR switches on. Observe the value of gate voltage at which switching occurs. Alternatively, set

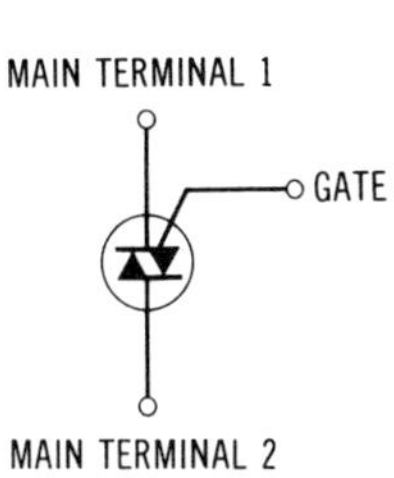

(*A*) *Schematic.*

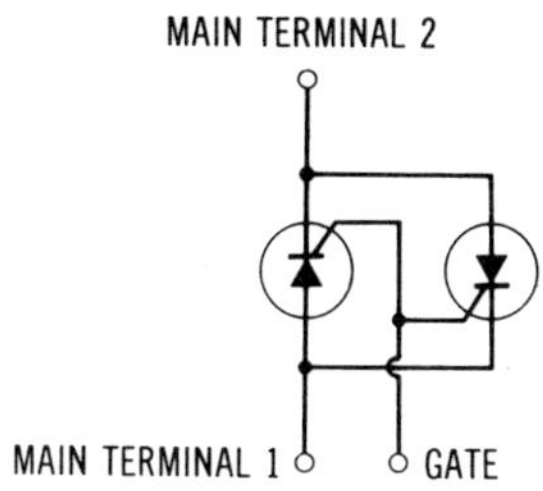

(*B*) *Equivalent circuit.*

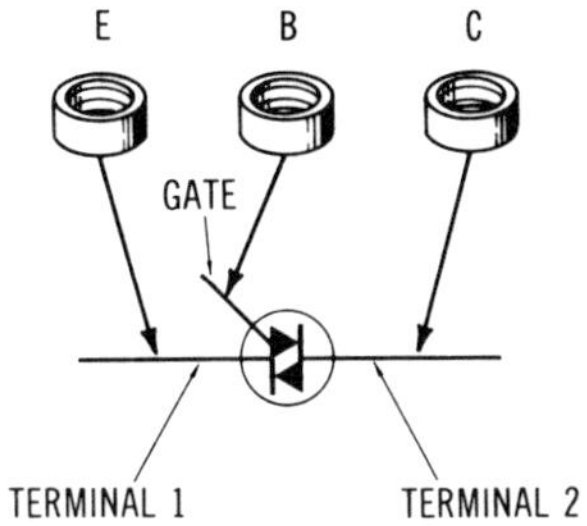

(*C*) *Connection to curve tracer.*

Fig. 5-25. A triac consists of two paralleled and oppositely polarized SCRs.

the dc bias supply to a specified gate voltage, and increase the sweep voltage until the SCR switches on. Observe the peak value of sweep voltage that is required for switching action.

Evaluation of Results: The turn-on point of an SCR depends on both the forward voltage and the value of gate voltage that are applied. As the gate voltage is increased, less forward voltage is required to switch the SCR, and vice versa. Thus, the SCR is evaluated for the bias voltage with respect to the sweep voltage required for switching action.

NOTE 5–14

Triac Operation

Triacs are four-layer pnpn devices that have the same characteristics in both directions; they are used in ac applications. As depicted in Fig. 5-25 a triac is the equivalent of two SCRs connected in parallel, but oriented in opposite directions. A triac has a main terminal 1, a main terminal 2, and a gate terminal. Triacs are tested in the same manner as SCRs except that forward tests are repeated to determine the characteristics in both directions. Note that there is no reverse blocking voltage measurement.

SECTION 6

Miscellaneous Applications

6–1. To Measure the Bandwidth of an LC Circuit With a Ringing Test

Equipment: Square-wave or pulse generator that provides fast rise, coil, and capacitor (fixed or variable).

Connections Required: Connect equipment as described in Fig. 6-1.

Procedure: Display the ringing waveform obtained with a chosen value of C, and count the number of peaks (cycles) in the ringing waveform from the 100 percent to the 37 percent amplitude point. Measure the ringing frequency.

Evaluation of Results: Multiply the number of cycles from the 100 percent to the 37 percent amplitude point by pi (3.14); this gives the approximate Q value of the LC circuit. Then, divide the ringing frequency by the Q value; this gives the approximate bandwidth of the LC circuit. As an illustration, suppose that the Q value of a coil is 50 and that the ringing frequency is 1 MHz. In turn, the bandwidth of the coil and its associated capacitance is approximately 20 kHz. Bandwidth is defined in this method as the number of cycles between the 0.707 voltage points on the frequency response curve. These 0.707 voltage points are also called the half-power points, or the −3 dB points.

NOTE 6–1

Resonant-Frequency Formulas

A damped sine wave such as shown in Fig. 6-1 has a frequency which is given approximately by the familiar resonant-frequency formula:

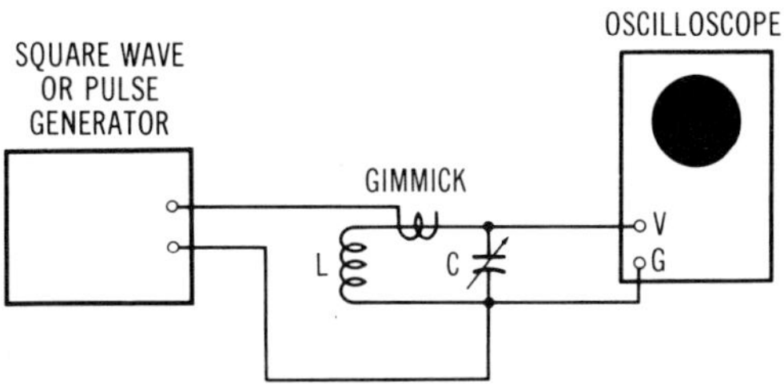

(*A*) *Test setup.*

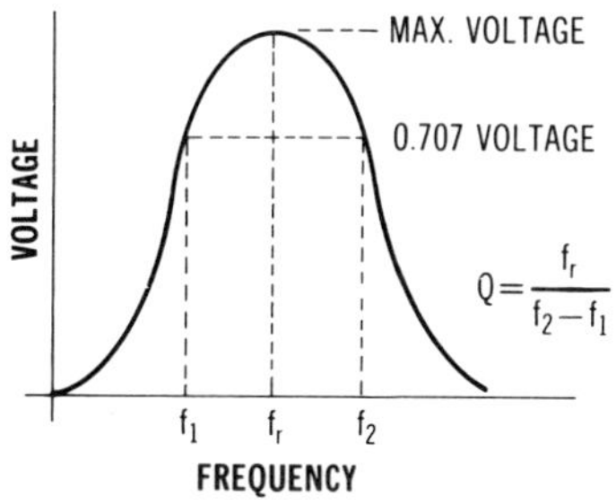

(*B*) *Computing Q.*

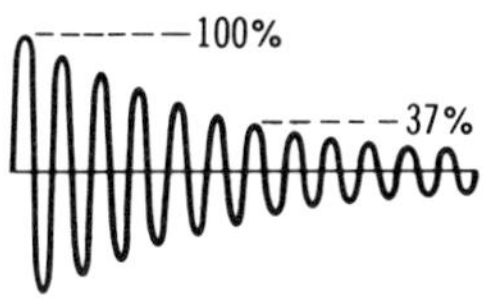

(*C*) *Pattern.*

Fig. 6-1. Determination of bandwidth for LC circuit.

$$f = \frac{1}{2\pi\sqrt{LC}}$$

However, the exact ringing frequency is formulated:

$$f = \frac{1}{2\pi}\sqrt{\frac{1}{LC} - \frac{R^2}{4L^2}}$$

We will find that when R has a small value, as in a color-subcarrier ringing-oscillator circuit, that the familiar resonant-frequency formula gives practically the same answer as the exact formula.

6–2. To Make a Ringing Test of an IF Transformer

Equipment: Square-wave or pulse generator with fast rise.

Connections Required: Connect equipment as shown in Fig. 6-3.

Procedure: Display the ringing waveform, and adjust the trimmer capacitors (or slugs) for maximum pattern amplitude; well-defined zero-beat crossings will be observed at maximum amplitude.

Evaluation of Results: In general, this is considered to be a comparative test, to weed out defective transformers. How-

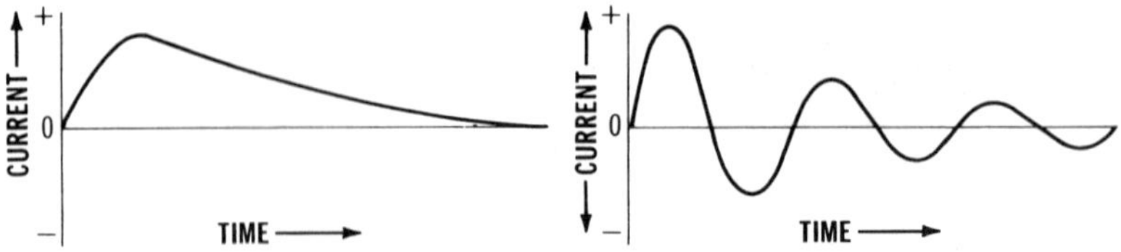

RESISTOR VOLTAGE WAVEFORM WHEN R IS LARGE — RESISTOR VOLTAGE WAVEFORM WHEN R IS SMALL

REDUCTION IN RINGING FREQUENCY CAUSED BY CIRCUIT RESISTANCE

RESISTANCE* (% of R_C)	REDUCTION IN FREQUENCY $\left(\% \text{ OF } \frac{1}{2\pi}\sqrt{1/LC}\right)$
0 (ZERO RESISTANCE)	0 $\quad f_0 = \frac{1}{2\pi\sqrt{LC}}$
10	.5
30	4.6
50	13.4
70	28.6
90	56.4
100	100.0 (NO OSCILLATION)

*R_C DENOTES THE CRITICAL RESISTANCE VALUE, OR $2\sqrt{L/C}$.

WHEN THE CIRCUIT RESISTANCE IS INCREASED TO R_C, THE RINGING FREQUENCY IS ZERO, AND THE RINGING WAVEFORM BECOMES A TRANSIENT SURGE. THE Q VALUE BECOMES ½ AT CRITICAL DAMPING.

CRITICAL DAMPING IS OF CONSIDERABLE IMPORTANCE IN SPEAKER OPERATION

Fig. 6-2. Relation of natural frequency (ringing frequency) to the driven resonant frequency, as in an amplifier load circuit.

ever, a few of the pattern characteristics are of interest. An overcoupled if transformer has two resonant frequencies (two hump frequencies) which result from the mutual inductance between the two windings. These hump frequencies beat together to form the ringing pattern. If we call these ringing frequencies f_1 and f_2, the beat (envelope) frequency will be $f_2 - f_1$. The center frequency of the if transformer, such as 456 kHz, 4.5 MHz, or 10.7 MHz, is equal to the ringing frequency, or to $(f_1 + f_2)/2$.

NOTE 6–2

Adjustment of Primary and Secondary Frequencies

If the primary and secondary windings are not tuned to the same frequency in the procedure of 6–2, the ringing pattern will display peaks

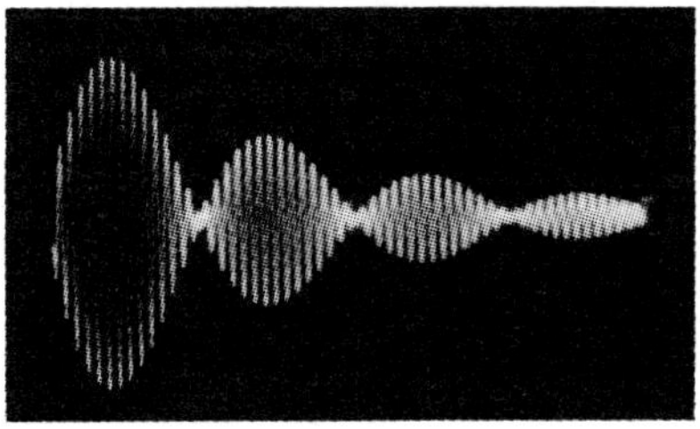

(*A*) *Typical ringing pattern.*

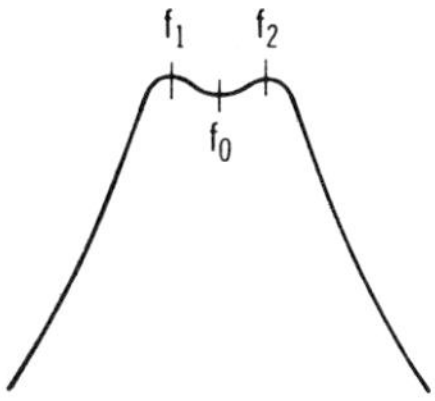

(*B*) *The two resonant frequencies are f_1 and f_2, and the ringing frequency is f_0.*

(*C*) *Equivalent circuit, showing mutual inductance L_M.*

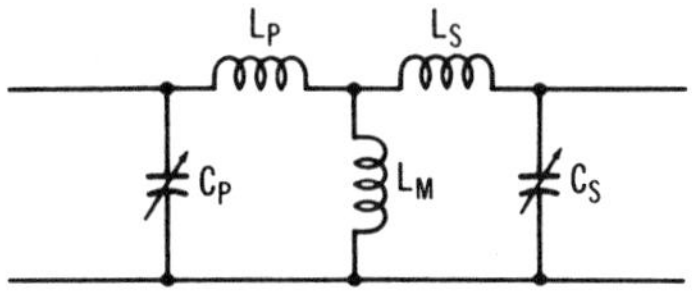

Fig. 6-3. Ringing test of an if transformer.

and valleys but will not show any zero (crossover) points. As the primary and secondary frequencies are adjusted more nearly equal, the valleys in the waveform approach the zero level. At widely different primary and secondary frequencies, the higher frequency may appear merely as a "ripple" along the lower frequency.

6–3. To Measure Percentage of Amplitude Modulation

Equipment: An am signal generator or other source of amplitude-modulated signal. Audio oscillator (if signal generator does not have audio output). Trapezoidal method requires an rf pickup loop and an RC coupling circuit, as shown in Fig. 6-4.

Connections Required: When the time-base display is used, connect the source of the am signal directly to the vertical-input channel of the scope. When the trapezoidal method is used, connect the pickup loop to the vertical-input channel of the scope, and connect the RC coupling circuit from the modulating voltage source to the horizontal-input channel of the scope.

Procedure: In the time-base method, observe the maximum and minimum peak voltages in the pattern. When the trapezoidal display is used, adjust the pickup coil orientation and the potentiometer for a trapezoidal pattern with proper layover;

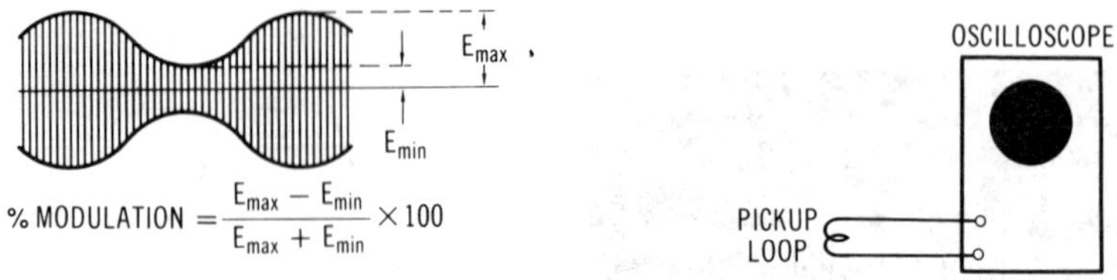

(*A*) *Time-base method.*

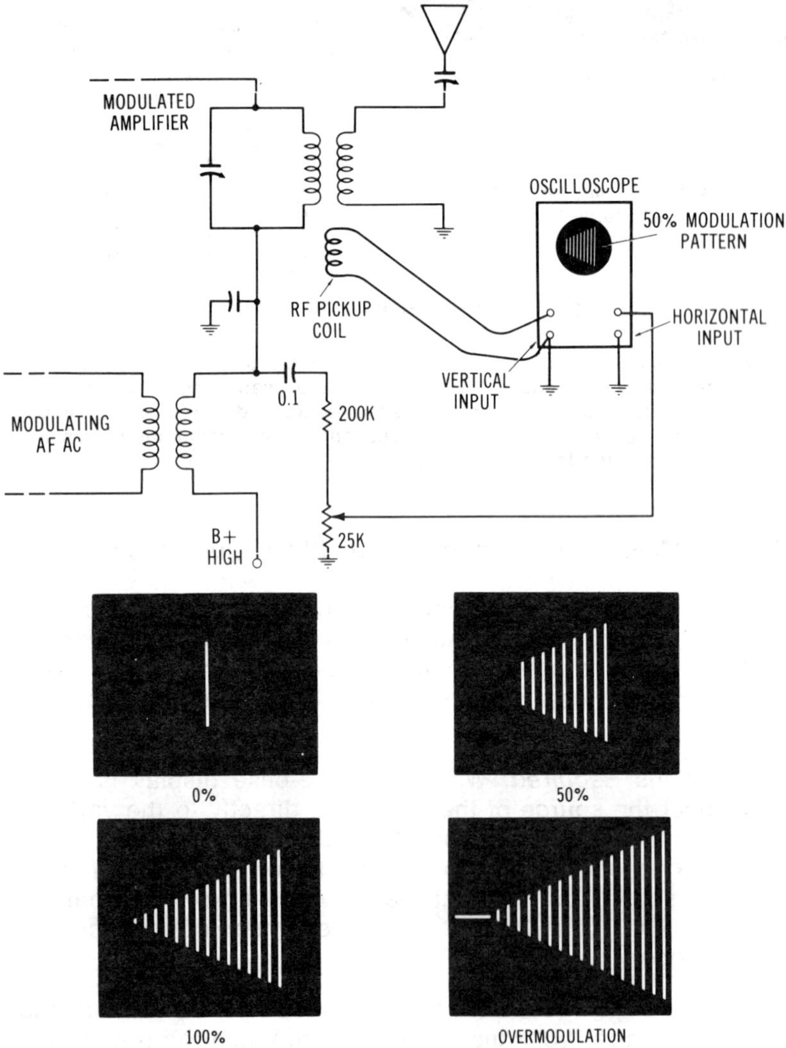

(*B*) *Trapezoidal method.*

Fig. 6-4. Measurement of percentage amplitude.

observe the relative vertical deflections at the ends of the trapezoidal pattern.

Evaluation of Results: Percentage of amplitude modulation is calculated as noted in the diagram.

NOTE 6–3

Distorted Modulation Patterns

Amplitude-modulation patterns may show various types of distortion. For example, the illustration in Fig. 6-5 exemplifies typical inherent distortion for an economy-type am generator. This is an example of nonsymmetrical and nonsinusoidal modulation. Residual frequency modulation is also present in the carrier, as a result of direct modulation of the rf oscillator. When trapezoidal displays are used, am distortion shows up curvatures or other distortions in the top and bottom of the trapezoid; the ends of the pattern may also be displaced vertically with respect to each other. Nonsymmetrical modulation causes a "double image" display, due to poor layover of the forward and reverse images.

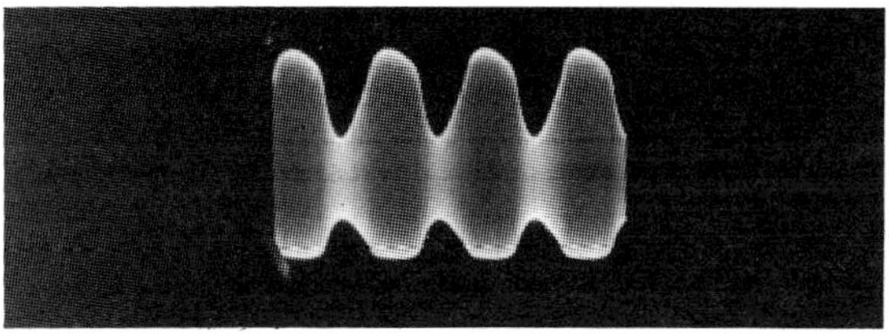

Fig. 6-5. Output signal from an economy-type am generator.

6–4. To Check the Operation of an Op-Amp Integrator

Equipment: Square-wave generator.

Connections Required: Connect output from square-wave generator to input of op-amp integrator. Connect output from integrator to vertical input channel of scope (Fig. 6-6).

Procedure: Observe output waveform over the rated repetition rate for the op amp.

Evaluation of Results: A precise triangular waveform is normally displayed on the scope screen. The useful range of the exemplified configuration is from 1 to 10 kHz, approximately.

NOTE 6–4

Op-Amp Characteristics

An operational amplifier (op amp) is a very high gain integrated-circuit amplifier that is ordinarily used with considerable negative feedback.

In turn, the op amp has very low distortion, high stability, and a voltage amplification that depends on the value of the feedback resistance or feedback impedance. An op amp has comparatively high input resistance (in the absence of feedback), and a low output resistance. When a large amount of negative feedback is used, as in the example of Fig. 6-4, the input impedance of the op amp becomes very low, and can be regarded as a virtual ground. Consequently, the effective input impedance of the configuration is equal to the series input resistance that is used, plus the internal resistance of the source. Note also that an op amp has two input terminals; the inverting input is designated (—), and the noninverting input is designated (+). In the example of Fig. 6-4, the noninverting input is grounded via the 39 K resistor at terminal 4, and the inverting input is driven via Z_r (the series input resistor, or weighting resistor). The weighting resistor determines the output/input voltage ratio, with all other things being equal.

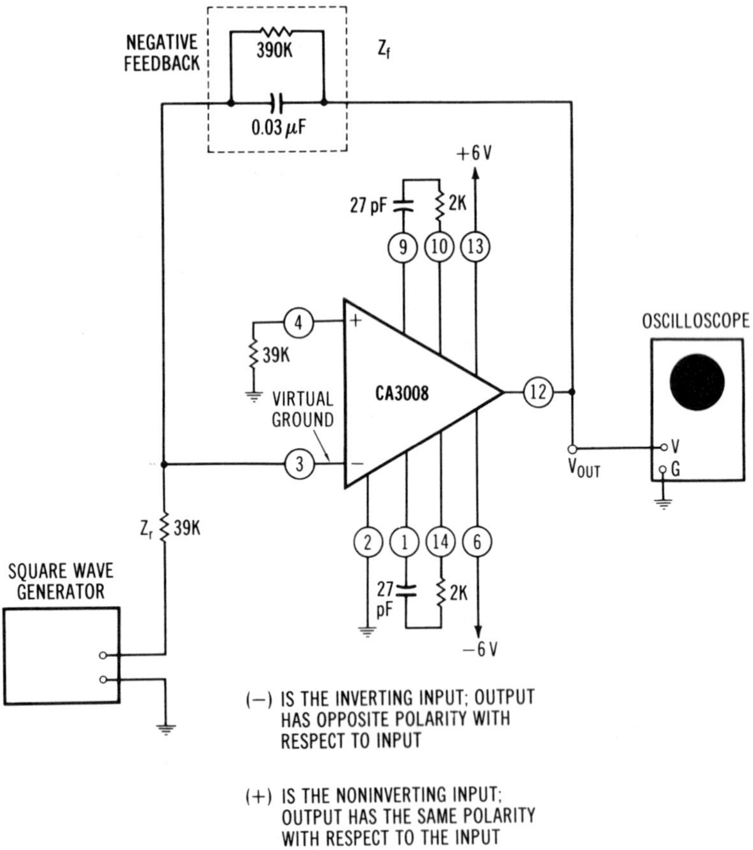

(A) Test setup.

Fig. 6-6. Checking the operation

6–5. To Check the Operation of an Op-Amp Differentiator

Equipment: Square-wave generator.

Connections Required: Connect test setup as shown in Fig. 6-7.

Procedure: Observe the output waveform from the differentiator.

Evaluation of Results: A very narrow pulse output is normally produced at each leading and trailing edge of the square-wave input. Note that the 51-ohm resistor Z_r in series with the input capacitor is used to minimize noise interference in the output (the resistor limits the rf gain of the op amp).

NOTE 6–5

Ideal Differentiator Action

An op-amp differentiator provides an output which more nearly approximates a true mathematical derivative than does an RC differentiator. An ideal differentiator would produce an output pulse with infinite amplitude and infinitesimal width. Similarly, an op-amp integrator produces an output waveform which approximates a true mathematical integral more nearly than does an RC integrator. In other words, RC differentiators and integrators provide exponential output waveforms (Fig. 6-8),

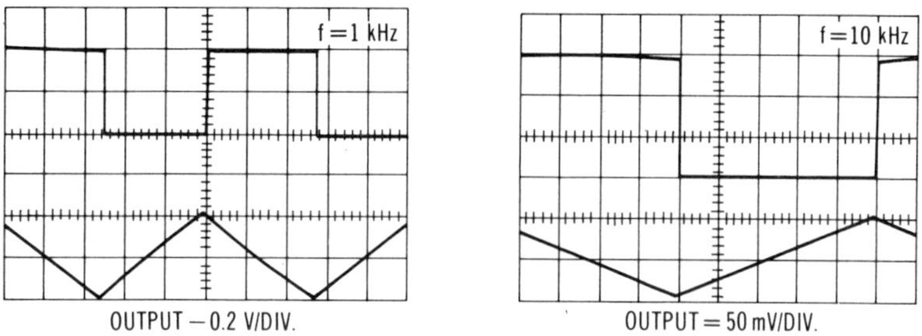

(*B*) *Normal operating waveforms.*

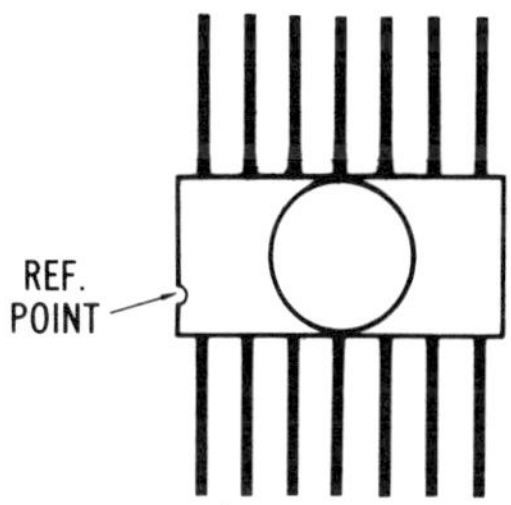

(*C*) *Typical package.*

of an op-amp integrator.

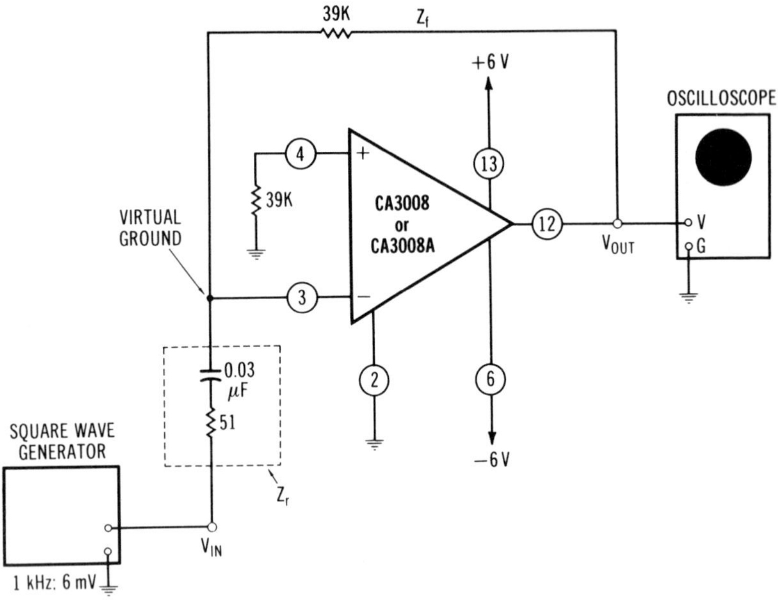

(*A*) *Test setup.*

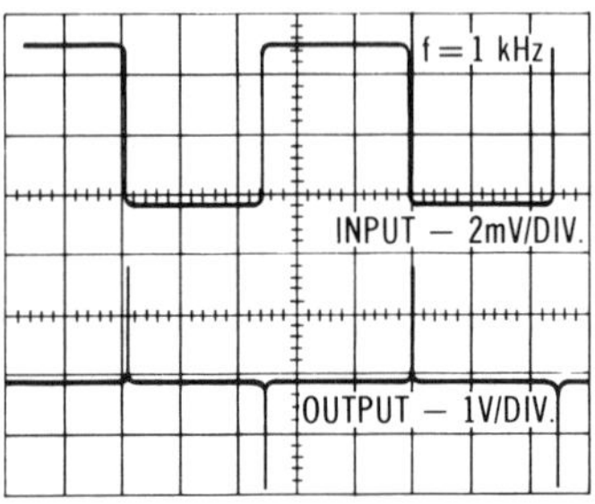

(*B*) *Normal input/output waveforms.*

Fig. 6-7. Checking the operation of an op-amp differentiator.

whereas op-amp differentiators and integrators provide good approximations to straight vertical lines and straight-sided triangular waveforms. (Op amps are used extensively in function generators and in analog computers.)

6–6. To Check the Operation of an Op-Amp Summer (Adder)

Equipment: Audio oscillator and square-wave generator.

Connections Required: Connect test setup as shown in Fig. 6-9.

Procedure: Adjust the audio oscillator output to a suitable level such as 90 mV p-p, and adjust square-wave generator output to a higher level, such as 180 mV p-p. Use different frequen-

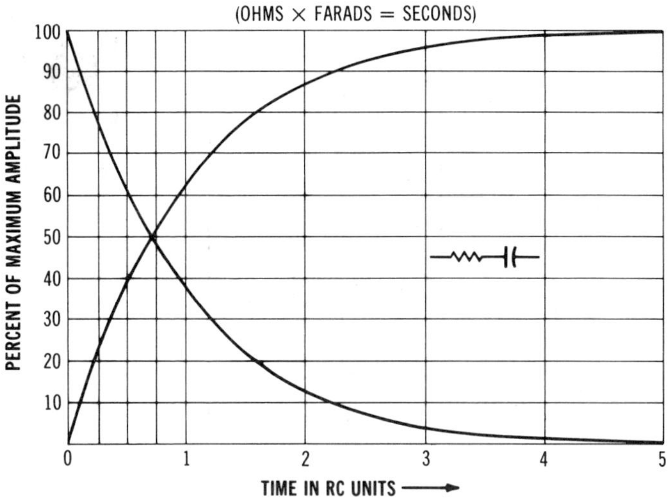

Fig. 6-8. RC differentiators and integrators produce exponential output waveforms.

cies in the two input signals, such as 1 kHz and 100 Hz. Observe output waveform.

Evaluation of Results: As depicted in the diagram, an op-amp summer will normally produce the sum of the two input waveforms, at an amplified level: Observe that 1K input (weighting) resistors are used in this example. Three, four, or more inputs, each with a 1K input resistor, may be used, if desired, because terminal 3 of the op amp is a virtual ground, due to the large amount of negative feedback that is utilized.

NOTE 6-6

Subtraction by Op-Amp Summer

An op-amp summer will add dc voltages, if desired, in the same manner as it adds ac voltages. Note that a summer operates as a subtracter, if one of its inputs is reversed in polarity. Thus, +2 V added to +3 V equals +5 V; on the other hand, −2 V added to +3 V equals +1 V. Or, +2 V added to −3 V equals −1 V. In the case of ac input voltages to a summer, the output waveform displays their sum when the two input waveforms are in phase with each other. On the other hand, the output waveform displays their difference when the two input waveforms are 180° out of phase with each other.

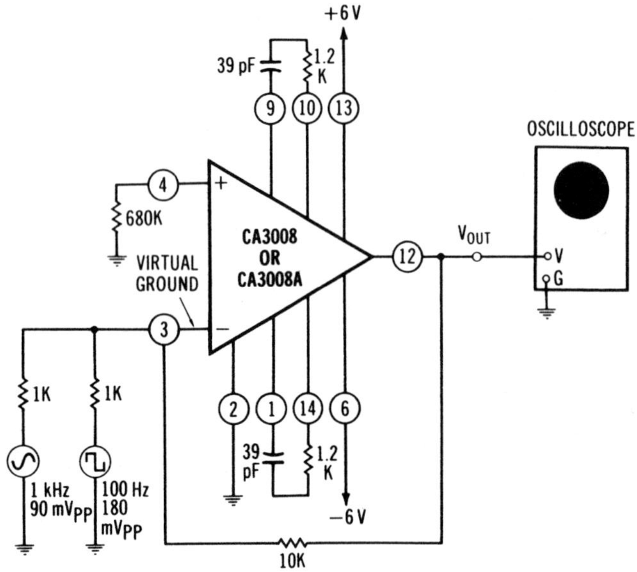

(*A*) *Test setup.*

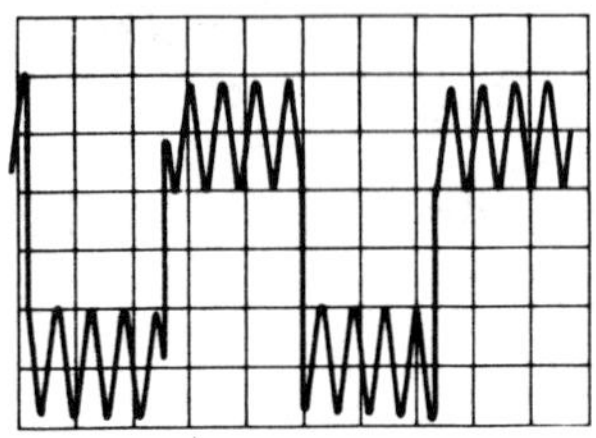

(*B*) *Normal output waveform.*

Fig. 6-9. Checking the operation of an op-amp summer (adder).

6–7. To Check the Operation of an Op-Amp Scaling Adder

Equipment: Audio oscillator and square-wave generator.

Connections Required: Connect test setup as shown in Fig. 6-10.

Procedure: Same as in 6–6.

Evaluation of Results: This op-amp adder employs a 1 K input resistor for the sine-wave signal, and a 2 K input resistor for the square-wave signal. Practically, the gain for each input is calculated as the sum of the feedback and input resis-

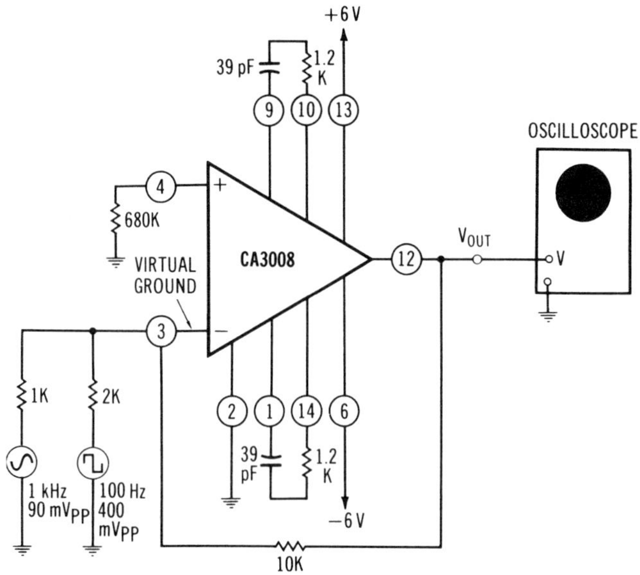

(A) *Test setup.*

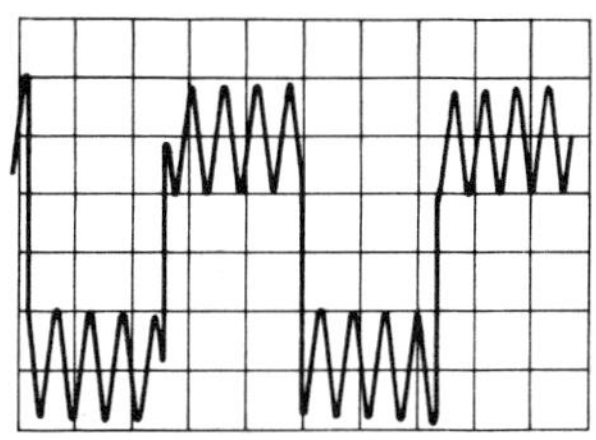

(B) *Normal output waveform.*

Fig. 6-10. Checking the operation of an op-amp scaling adder.

tances, divided by the input resistance. Thus, the gain is normally 11 times for the sine-wave signal, and 6 times for the square-wave signal. In turn, the output waveform normally appears as illustrated. The same test principles apply to more elaborate scaling adders that have more than two inputs.

SECTION 7

Oscilloscope Performance Checks

7–1. To Check the Accuracy of a Calibrated Time Base

Equipment: Good-quality audio oscillator and lab-type am generator.

Connections Required: Connect the output from the audio oscillator or from the signal generator to the vertical-input channel of the scope.

Procedure: Set the audio oscillator (or am generator) to various low, medium, and high frequencies; observe the amounts of horizontal deflection occupied by one cycle of the displayed waveform at various settings of the time/cm control.

Evaluation of Results: The period of the displayed waveform normally is equal to 1/f, where f is the applied test frequency. For example, in Fig. 7-1 the test frequency is 2500 Hz; the corresponding period is 400 μs. In turn, if the time/cm control is set for 100 μs/cm, the displayed waveform will normally occupy four horizontal divisions of the screen.

7–2. To Check the Accuracy of a Calibrated Vertical Attenuator

Equipment: Audio oscillator and audio voltmeter.

Connections Required: Connect output from audio oscillator to vertical input terminals of scope and to input terminals of the audio voltmeter. Test signal may be attenuated by 100-to-1

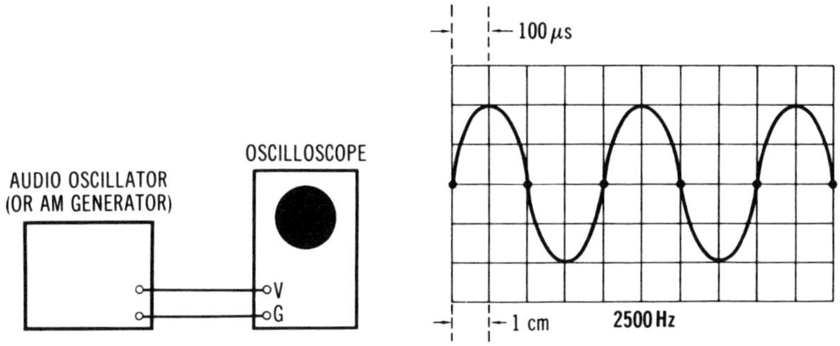

(*A*) *Test setup.* (*B*) *Typical screen display.*

Fig. 7-1. Checking the accuracy of a calibrated time base.

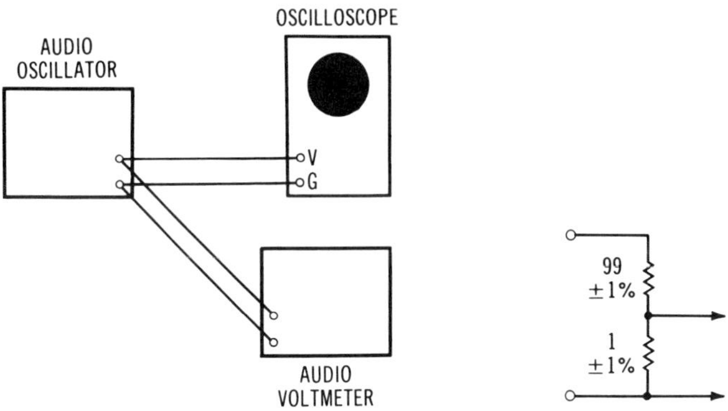

(*A*) *Test setup.* (*B*) *100-to-1 pad to obtain low-level test voltages.*

Fig. 7-2. Checking the accuracy of a calibrated vertical attenuator.

resistive pad, if required to obtain a suitably low-level signal (see Fig. 7-2).

Procedure: Check each step of the vertical-input attenuator and compare the peak-to-peak voltage indicated on the scope screen with the reading of the audio voltmeter. (Audio voltmeter may indicate in rms values, or both rms and peak-to-peak values; the peak-to-peak value is equal to 2.83 times the rms value.)

Evaluation of Results: The peak-to-peak voltages indicated on the scope screen should agree within reasonable tolerance with the peak-to-peak voltages indicated by the audio voltmeter. Note that the scope is likely to have an internal calibrating voltage; if so, this provides a useful cross-check.

7–3. To Check the Vertical and Horizontal Amplifiers for Linearity

Equipment: Audio oscillator.

Connections Required: Apply output from audio oscillator to vertical and horizontal input channels of scope as shown in Fig. 7-3.

Procedure: Adjust horizontal and vertical gain controls for a pattern that extends to the edges of the screen. However, do not exceed the linearity limits.

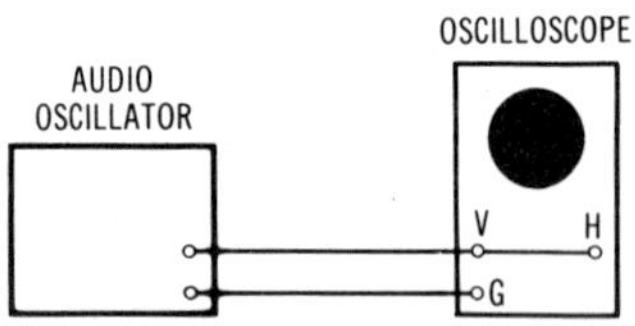

(*A*) *Test setup.*

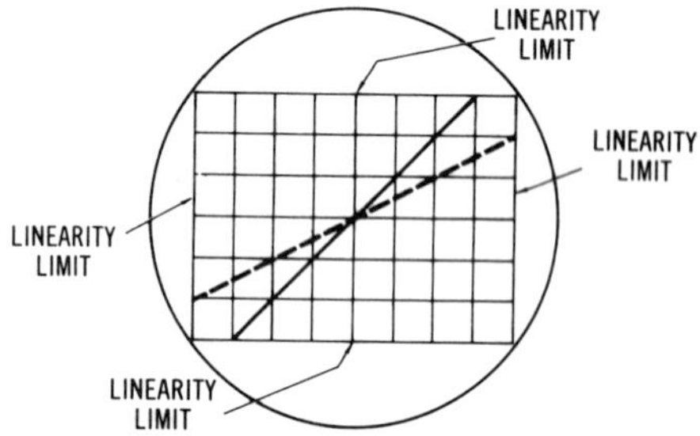

(*B*) *Normal display.*

Fig. 7-3. Checking the vertical and horizontal amplifiers for linearity.

Evaluation of Results: Diagonal lines are displayed on the screen; the slope of a displayed line depends upon the relative vertical and horizontal gain settings. In any case, a line that is within the linearity limits of the screen should be precisely straight. If curvature is present, it is indicated that the vertical amplifier, the horizontal amplifier, or both, is/are nonlinear. The relative amount of curvature changes as the gain-control settings are varied.

7–4. To Check the Phase Accuracy of a Dual-Trace Oscilloscope

Equipment: An am signal generator.

Connections Required: Connect output from generator to both the A and the B vertical-input channels, as shown in Fig. 7-4.

Procedure: Vary the test frequency up to the rated high-frequency limit of the scope, and observe the dual-trace pattern.

(A) Test setup.

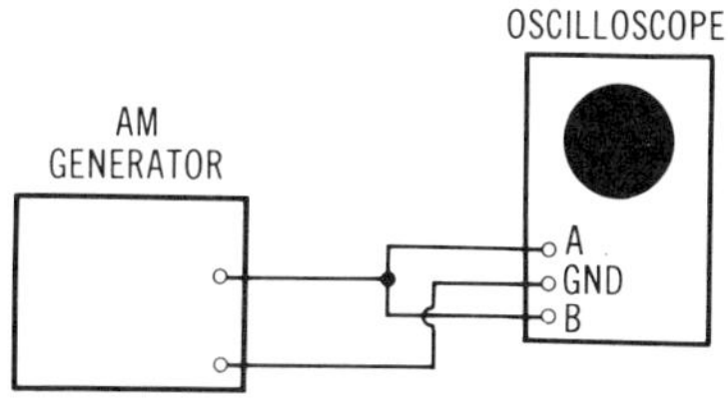

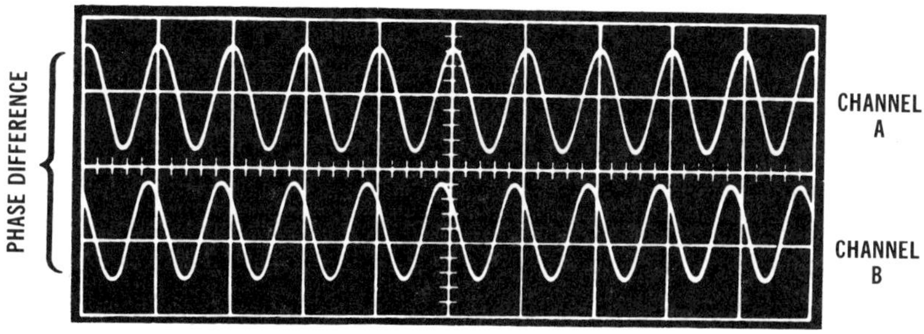

(B) Pattern with phase error present.

Fig. 7-4. Checking a dual-trace scope for A/B phase shift.

Evaluation of Results: The two patterns are normally precisely in phase with each other. If a phase error is present, it will become more evident at high frequencies. In such a case, the phase error may be noted and taken into account.

7–5. To Check a Vertical Amplifier for Overload Recovery

Equipment: An am generator and pulse generator.

Connections Required: Connect outputs from generators to vertical-input terminals of scope, as shown in Fig. 7-5.

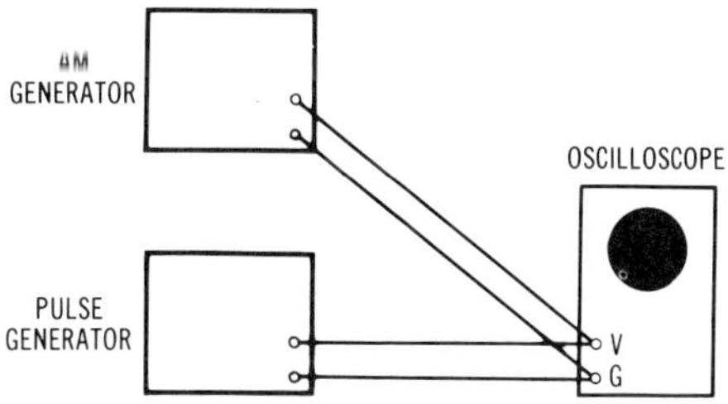

(A) Test setup.

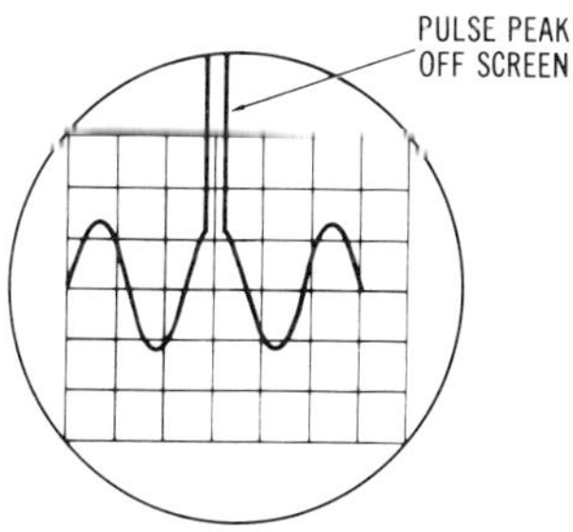

(B) Pattern.

Fig. 7-5. Checking the overload recovery capability of an oscilloscope.

Procedure: Synchronize the pulse generator from the signal generator, or carefully adjust the am generator frequency to make the sine wave "stand still" on the screen. Operate the am generator at a higher frequency than the pulse generator. Display the sine-wave pattern; then advance the pulse-generator output until the pulse component is off-screen vertically.

Evaluation of Results: The sine-wave pattern should not become distorted as the pulse component is increased into the overload region of the vertical amplifier. Note, however, that some lab-type scopes are designed for better overload recovery than others. When the overload recovery capability is good, a waveform may be expanded vertically as much as may be desired.

7–6. To Check a Differential-Input Scope for Common-Mode Rejection

Equipment: Square-wave generator.

Connections Required: Connect output from generator to one of the differential-input terminals and ground. Then, connect output from generator to both of the differential-input terminals and ground (Fig. 7-6).

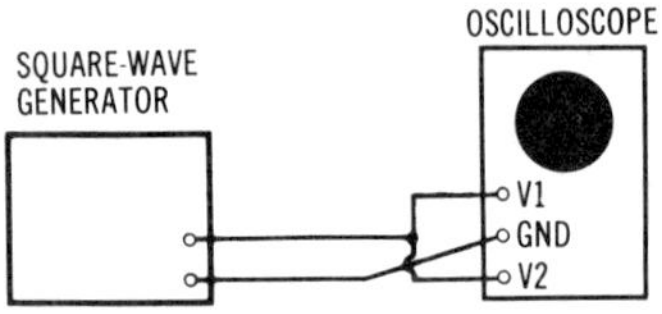

(*A*) *Test setup.*

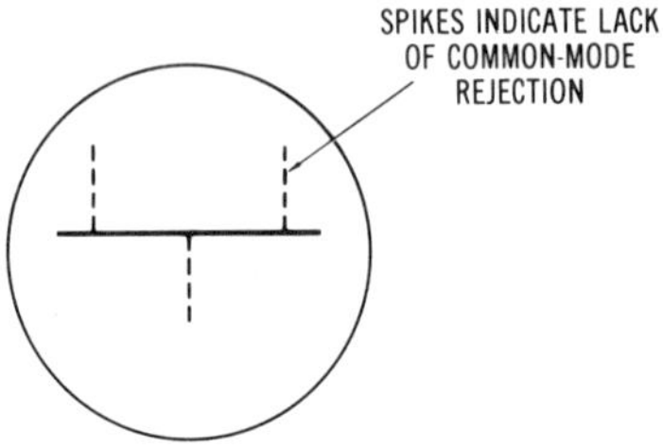

(*B*) *Pattern.*

Fig. 7-6. Checking a differential-input scope for common-mode rejection.

Procedure: With the square-wave signal driving one of the differential vertical-input terminals, adjust the corresponding vertical attenuator for full-screen deflection. Then, set the other vertical attenuator to the same level. Apply the square-wave signal to both of the differential vertical-input terminals, and observe the screen pattern (if any).

Evaluation of Results: Only a straight horizontal line will be displayed if common-mode rejection is complete. In practice, small irregularities may appear along the horizontal trace. If tall spikes appear in the pattern, common-mode rejection is faulty.

7–7. To Check the Response of an Oscilloscope to Line-Voltage Fluctuation

Equipment: Variac or equivalent, and audio oscillator.

Connections Required: Connect power cord of oscilloscope in series with the Variac; connect output from audio oscillator to vertical-input channel of scope.

Procedure: Display the audio sine-wave pattern on the scope screen. Vary the output voltage from the Variac over the rated power-line voltage range for the scope. Observe pattern changes (if any) while varying the line voltage slowly, and while varying it rapidly.

Evaluation of Results: Virtually no change in pattern size, brightness, or stability will occur in a scope that has complete voltage-regulation circuitry. On the other hand, an economy-type design may be susceptible to pattern jumping and/or displacement, loss of sync, change in pattern size, and change in pattern brightness as the line voltage is varied.

7–8. To Determine Whether an Oscilloscope Radiates RF Energy

Equipment: An am radio receiver, signal generator.

Connections Required: Connect output from signal generator to vertical-input channel of scope.

Procedure: Place radio receiver in vicinity of scope. Apply a 1-MHz signal to the scope, and display the waveform on high-speed sweep, such as 0.04 μs/cm.

Evaluation of Results: Tune the radio receiver to a station in the midband region, and vary the generator frequency. If the scope is radiating rf energy, a heterodyne squeal will be heard from the speaker. If the receiver is tuned through the am broadcast band, a rasping type of interference may be found at one or more regions of the band. *Some scopes have a comparatively high level of rf radiation, and the condition cannot be classified as a fault. In such a case, the operator should keep the radiation field in mind when making tests on nearby high-impedance high-frequency circuitry.*

7–9. To Determine Whether an Oscilloscope Case Is "Hot"

Equipment: An ac voltmeter and 100K resistor.

Connections Required: Connect ground terminal of scope to cold-water pipe via the 100K resistor. Connect ac voltmeter across the resistor.

Procedure: Turn scope "on" and observe meter reading (if any).

Evaluation of Results: The meter should read virtually zero, when the power switch of the scope is "off" or "on." A substantial voltage reading indicates that there is excessive power-line voltage gaining access to the scope case.

7–10. To Check the Immunity of an Oscilloscope to External Fields

Equipment: Radio transmitter operating within the rated frequency range of the scope.

Connections Required: Short-circuit the scope's vertical-input terminals.

Procedure: Place the scope in the vicinity of the transmitter. Operate the scope without any applied signal.

Evaluation of Results: If the scope is immune to external rf fields, no pattern will be displayed when the transmitter is operating. On the other hand, if the radiation field from the transmitter finds entry into the scope circuitry, the transmitter signal will be displayed on the screen. Note that the transmitter signal may gain entry via the power line, or because of incomplete shielding of the scope circuitry. It is sometimes helpful to ground the case of the scope to a cold-water pipe.

7–11. To Determine the Temperature Response of an Oscilloscope

Equipment: An am signal generator.

Connections Required: Connect output from am generator to vertical-input channel of scope.

Procedure: Observe pattern display when room temperature is chilly; then, turn on room heater and raise the ambient temperature to 90°. Observe pattern change (if any) as room temperature rises.

Evaluation of Results: A well-designed oscilloscope is virtually immune to room-temperature change. However, an economy-type scope may show substantial change in gain and/or stability with variation in room temperature. *To eliminate possible changes in output from the signal generator when the room temperature changes, the generator may be located in another room.*

Index